Seabrook Station

Henry F. Bedford

Seabrook Station

*Citizen Politics
and Nuclear Power*

The University of
Massachusetts Press
Amherst

Copyright © 1990 by
Henry F. Bedford
All rights reserved
Printed in the United States of America
LC 89–77123
ISBN 0–87023–711–X
Designed by Edith Kearney
Set in Linotron Electra by Keystone Typesetting, Inc.
Printed and bound by Thomson-Shore

Library of Congress Cataloging-in-Publication Data

Bedford, Henry F.
    Seabrook Station : citizen politics and nuclear power / Henry F.
Bedford.
        p.   cm.
    Includes bibliographical references.
    ISBN 0–87023–711–X (alk. paper).
    1. Seabrook Nuclear Power Plant (N. H.)   I. Title.
TK1344.N4B43 1990
363.17'99'097426—dc20   89–77123
                                                              CIP

British Library Cataloguing in Publication data are available.

For HJ, HO, and M.

May they do nuclear power,
among other things,
more competently.

Contents

Preface: *"The reactor is critical"*   ix

Abbreviations   xv

Chronology   xvii

1 Introduction: *"What can you do with a dead dinosaur?"*   3

2 The Environment: *"Brick by brick, pipe by pipe, fish by fish."*   31

3 The Opposition: *"No, No, No."*   64

4 Money and Management: *"Teeter[ing] precariously on the brink of financial disaster."*   94

5 Emergency Planning: *"Why make umbrellas if it's not going to rain?"*   125

6 Conclusion: *"A $5 billion mess."*   162

Epilogue   200

Notes   203

Index   217

Illustrations follow page 100

Preface

"The reactor is critical."

This is a book about people.

It is also about power, and not just about the power nuclear fission produces at an electrical generating station. It is about political power as well and the power derived from status and access and information and money.

And it is about fear. Fear of radiation and accidental catastrophe, of course. But also fear of losing an opportunity, a job, a reputation, an investment; of losing a struggle of such duration that the rivalry itself sometimes obscured its causes; of losing the amenities of a pleasant and familiar environment; of losing control of communities, neighborhoods, and lives.

To stake too pretentious a claim, it is also about the United States in the last third of the twentieth century, as revealed in the effort to construct, license, and operate a nuclear power plant at Seabrook, New Hampshire. That prolonged, expensive, sometimes bitter battle was an epic from which few emerged with honor and none with victory. Management performed ineptly every managerial task; government could not furnish prompt or persuasive resolutions of disputes and therefore governed badly; opponents could not prevent results they predicted would be disastrous. For the uninvolved public, it was a tale of unrelieved incompetence.

Unskilled management, bad government, failed grass-roots initiatives, an apathetic public, and pervasive incompetence were not unique to coastal New Hampshire or to the nuclear industry in these years. In some respects, the unfinished and rusting second Seabrook reactor, an appropriate symbol of the nation's nuclear policy, might have represented other blighted national aspirations equally well: the balanced budget, democracy in Central America, the end of homelessness or drug abuse. The controversy over Seabrook Station coincided with America's discovery that green was not the only hue in the cosmic traffic signal. For some, the light changed at a Memphis motel, a Watergate office, or a hamlet in Vietnam; for some, at Seabrook Station.

This Seabrook saga centers on the relationship of citizens and government. Electoral politics is not the main event, although the election of a couple of New Hampshire governors was important and that of George Bush perhaps decisive. But nuclear politics is not the politics learned in civics classes. Legislatures are involved, but they are less important than agencies and bureaucrats; the judicial system is invoked, but it is less relevant than administrative judges. Nuclear licensing is settled in the executive branch of government, according to rules made, for the most part, by the executive branch itself. The public is given a voice but rarely an attentive ear, and nobody asks for a show of hands. It is not, in the conventional sense, a democratic procedure.

Nor, as even its defenders admit, does it work very well. The most charitable description of Seabrook's licensing noted the expense and slow pace. Fifteen years, a library full of documents, and millions of dollars were only part of the evidence that led others to interpret the process as fragmented, irrational, and rigged. It focuses on technical and legal detail, an emphasis that gives every advantage to those with the deepest pockets and most information. What seems to the public a local problem, susceptible to ordinary intelligence and local rules, retreats through a governmental maze that appears to the uninitiated conceived to make simple matters complicated, to baffle, conceal, and frustrate. That perception may discourage, or even alienate; it may also inspire determined anger of the sort that sustained Seabrook's opponents through two decades, during which their command of detail, their legal sophistication, their numbers, and their political leverage grew.

They learned, in other words, something of the style of those who wanted to complete and operate Seabrook Station. Many who fought the plant in the 1970s believed fervently that people, even individuals, could make a difference; those who built it relied on systems: quality assurance systems, cooling water systems, management systems, monitoring systems, alarm systems. Boards and committees and agencies and commissions made decisions, systems did the work, and the public relations staff made everything look good. Actions had no face, no one took responsibility, and nobody erred. People who practiced civil disobedience at Seabrook's gates believed individuals answerable for their actions; people behind the gates believed systems made individuals irrelevant. The struggles differed, but that clash of perspectives was at the root of controversies in other corners of the United States as well.

"Attention in the plant," ran the laconic announcement from the loudspeaker late in June 1989; "the reactor is critical." In nukespeak shorthand, the adjective meant that fission had begun, and it was somehow appropriate that the message came from a faceless voice. The word "critical" had other connotations too, several of which were relevant. The plant's operation, its owners said, was "critical" to New England's energy supply and continued economic growth, not to mention the economic health of corporations and the people who owned them. Opponents, of course, were "critical" of a regulatory process that had subordinated the common sense, the wishes, and the democratic faith of ordinary people to privilege. Study of recent environmental politics, Samuel Hays has written, provides perspective on "emerging public values and objectives" and on "social change and governance in America." Seabrook Station has been my vantage point on American life in our time.

No one who has observed nearly two decades of debate over Seabrook Station lacks opinions about the controversy. Some of mine are no doubt evident. My intent, however, is to explain events rather than to participate belatedly in them. The explanation is partly contextual; local occurrences in the later twentieth century are never entirely local. Yet I have only alluded to such national developments as the growth of the environmental movement or the deterioration of American technological mastery, for instance, both of which are relevant, in order to clarify the Seabrook story first. Readers may, and unquestionably will, weave their own context around my work and supply their own interpretations.

My narrative rests on the public record. Nuclear regulation generates tons of documents of astonishing variety: volumes of oceanographic research, financial records, tens of thousands of pages of stenographic transcripts, legal motions, briefs, and letters. The Nuclear Regulatory Commission (NRC) also adds the relevant proceedings of other governmental bodies, such as the Environmental Protection Agency and the Federal Emergency Management Agency to the Seabrook files. The bankruptcy court in Manchester and the Securities and Exchange Commission hold additional financial data. Various state agencies in several New England states also have important files, copies of most of which find their way to the NRC. Thus this book depends in the main on material held in that agency's public document rooms in Washington or in the Exeter Public Library in Exeter, New Hampshire.

As a general rule, nuclear power has attracted the attention of the nation's enterprising journalists only during crises: the demonstrations of the 1970s, accidents at Three Mile Island or Chernobyl, the financial perils of utilities, the mismanagement of the nation's production of nuclear weapons. The regional press—the *Boston Globe, Manchester Union-Leader, Concord Monitor, Foster's Daily Democrat*, and *Hampton Union*, for example—has become more attentive to the issue as it has engaged more of the area's readers.

Opponents of nuclear power tend to collect information that documents their conviction, and several of them have shared with me this material, most of which had been published or was otherwise publicly available. Portions of the organizational records of the Seacoast Anti-Pollution League and of the Clamshell Alliance are the least accessible of the documentary evidence I have used. The Employee's Legal Project permitted me to examine the affidavits of Seabrook workers that furnished the basis for the project's allegations of drug abuse and shoddy craftsmanship. New Hampshire Yankee gave me access to consultants' reports and other corporate documents that were not generally available. The advantage these favors conferred was convenience; the narrative would not have been substantially altered without them.

That realization made easier a decision to curtail an ambitious plan for interviews, which seemed likely in any case to convey a one-sided antinuclear message. Opponents of the plant, who often felt no one was hearing them, volunteered readily both information and opinion; the more interesting portions of these conversations, however accurate, could not always be verified. Nor could I check these reminiscences against accounts of the plant's proponents, for officials of the Public Service Company of New Hampshire refused to speak with me. In the case of the legal team representing the company, refusal seemed proper; corporate officials ignored not only my requests but those of their colleagues at New Hampshire Yankee. Questions remain that perhaps interviews might have settled: the connections among politicians, regulators, investment bankers, and corporate officials, for example. But I grew increasingly doubtful that I could effectively supplement the public written record with verifiable oral evidence.

Donald Stever, the author of the only book on Seabrook Station, *Seabrook and the Nuclear Regulatory Commission* (Hanover, N.H.:

University Press of New England, 1980), was himself a participant in early Seabrook regulation and also relies almost exclusively on the public record. An assistant attorney general of New Hampshire, Stever represented the public in the state's hearings and the state in early NRC litigation. Other published material—of greater or less reliability—that bears on the subject of the construction and licensing of nuclear power plants is cited in the Notes. Samuel P. Hays, *Beauty, Health, and Permanence* (Cambridge: Cambridge University Press, 1987), analyzes recent environmental politics and thus provides context for this aspect of the study.

I have not invariably acknowledged in the text the authors of documents I have used, an oversight that is not always avoidable because of lack of identification; I acknowledge here witnesses, judges, lawyers, bureaucrats, and the often unnamed ordinary people who speak on these pages. Several of them have helped me in ways that ought to be acknowledged specifically. Herbert Moyer let me see his files on the early years of the Seacoast Anti-Pollution League, and Guy Chichester loaned me material on the Clamshell Alliance. Sharon Tracy provided personal recollection and access to material collected by the Employee's Legal Project. Stephen Comley sent me his voluminous clipping file. Robert Backus, an able advocate of the antinuclear cause, gave me remarkably detached counsel about the controversy. Barbara James organized a conference that provided an audience for my early thoughts on the subject and then served as my unpaid and inadequately thanked research assistant. Ron Sher, of the public relations staff of New Hampshire Yankee, was also an unofficial research assistant. He provided photographs, detailed chronologies, corporate reports, and access that would have been much more difficult without his support. When Judge Ivan Smith cleared a hearing after disrupting protests, for instance, Ron decided I might become a consultant to New Hampshire Yankee for a few hours and thus hear the rest of the day's testimony. David Schwab, the coordinator of Seabrook Station's Education Center, was also helpful. None of these people asked my views or sought to influence my interpretation of evidence they provided.

The staff of the Exeter Public Library has struggled for years to find space for the torrent of Seabrook documents and to make them available to readers. A new building has eased that task for Pam Gjettum, Lee Perkins, and the other librarians, all of whom assisted my research with unobtrusive skill. The staff at the public document room at the NRC attended promptly to my requests for catalogs and copies. Jane Boesch, the reference librarian at Phillips Exeter Academy, and the staff at libraries of the universities of New Hampshire and New Mexico also located material for me.

Pam Gjettum not only bent the regulations of the Exeter Public Library for me but also read and commented on several chapters of the manuscript. Richard D. Schubart, a former colleague and an encouraging editor, gave me a detailed critique of an early version of about half the book. Herbert Moyer read and corrected a draft of the chapter on Seabrook's opponents, and my son, Henry, applied his financial acumen to the chapter on management. My wife, whose involvement with Seabrook Station antedates my own, served as my first and most candid reader. Richard Martin and Bruce Wilcox, of the University of Massachusetts Press, provided encouragement and editorial expertise. Donna Griswold, Donna Arsenault, and Bette Parent instructed the miraculous machines that convert amateur typing into professional copy.

While I was dean of admission at Amherst College, Peter Pouncey and Richard Fink supported my desire to undertake this investigation and made possible my inattention to the office that research in New Hampshire necessitated. The staff of the Wilson Admission Center—receptionists, secretaries, assistant deans, and student guides—had to pick up the resulting slack, which they did with commendable skill. They also maintained a considerate and tactful silence about the things I left undone when the book got in the way. Perhaps publication will demonstrate that I was not simply sitting on the beach.

Abbreviations

ACRS	Advisory Committee on Reactor Safeguards
AEC	Atomic Energy Commission
AFUDC	Allowance for funds used during construction
ALAB	Atomic Safety and Licensing Appeal Board
ASLB	Atomic Safety and Licensing Board
CWIP	Construction work in progress
DES	Draft Environmental Statement
ELP	Employee's Legal Project
EPA	Environmental Protection Agency
EPZ	Emergency planning zone
FEMA	Federal Emergency Management Agency
FERC	Federal Energy Regulatory Commission
FES	Final Environmental Statement
FPC	Federal Power Commission
LPZ	Low population zone
MAC	Management Analysis Corporation
MIT	Massachusetts Institute of Technology
MMWEC	Massachusetts Municipal Wholesale Electric Company
NAI	Normandeau Associates, Inc.
NECNP	New England Coalition on Nuclear Pollution
NEPA	National Environmental Policy Act
NHRERP	New Hampshire Radiological Emergency Response Plan
NHY	New Hampshire Yankee
NRC	Nuclear Regulatory Commission
PSNH	Public Service Company of New Hampshire
PUC	Public Utilities Commission
RAC	Regional Assistance Committee

SAPL	Seacoast Anti-Pollution League
SEC	Securities and Exchange Commission
SPNHF	Society for the Protection of New Hampshire Forests
TMI	Three Mile Island
UE&C	United Engineers and Constructors, Inc.
YAEC	Yankee Atomic Energy Company

Chronology

1972

February PSNH applies to New Hampshire Site Evaluation Committee for approval of Seabrook site for a nuclear generating station.

June PSNH and eight other New England utilities sign memorandum of joint ownership.
Site hearings begin.

1973

March PSNH applies to AEC for construction permit; rejected in May; formally docketed in July.

1974

January Site Evaluation Committee approves Seabrook location.

April Draft Environmental Statement submitted; Final Environmental Statement issued in December.

1975

January Proposed date for start of construction; delayed until August 1976.
AEC replaced by NRC.

May ASLB hearings on safety and environmental issues begin; end in November.

October EPA approves tunnel location for once-through cooling.

1976

June	EPA approves design of cooling system. ASLB authorizes issuing of construction permit.
July	Construction permit issued.
August	Demonstrators attempt to block start of construction; 194 arrests made in two incidents.
November	EPA revokes approval of once-through cooling.
December	Court of Appeals denies petition to halt construction.

1977

April	NRC suspends construction permit pending resolution of environmental questions by EPA. 1,400 demonstrators arrested.
June	EPA reverses finding of November 1976 and again approves cooling system.
July	Construction permit reinstated; construction resumes on August 1.

1978

February	Court of Appeals reverses EPA decision of June 1977; orders the agency to conduct further hearings.
April	ALAB orders new hearings on construction but permits work to continue at site.
May	New Hampshire PUC permits inclusion in the rate base of charges for construction work in progress.
June	Clamshell Alliance converts planned demonstration to alternative energy fair. NRC orders construction suspended as of July 21 pending EPA determination that cooling system meets regulatory requirements.
August	EPA again concludes that cooling system is acceptable. Construction permit reinstated.
November	Hugh Gallen defeats Meldrim Thomson in New Hampshire gubernatorial election.

1979

March	PSNH board votes to reduce corporate share of Seabrook to 28 percent.

Accident at Three Mile Island.

May New Hampshire legislature prohibits inclusion of CWIP in rate base.

October Approximately 500 demonstrators arrested.

1980

March Unable to reduce its share of Seabrook below 35 percent, PSNH slows expenditure and reduces the work force at the site by about half.

June SAPL asks review of construction permit because of absence of emergency plans.

October NRC and FEMA issue new criteria for emergency planning.

1981

March Massachusetts asks suspension of construction until emergency plans have been devised and approved.

July Director of licensing denies intervenor motions for suspension of construction until emergency planning requirements are met.

1982

July PUC orders that PSNH's future borrowing not be used for completion of second reactor; overturned by New Hampshire Supreme Court in December.

1983

March–April Several minority partners announce opposition to management's announced plans to complete second reactor.

May Joint owners vote not to cancel or delay second reactor, in spite of escalating costs and slipping construction schedule.

August ASLB hearings on safety and emergency planning begin.

September Joint owners vote to reduce to "lowest feasible level" construction expenditure on second reactor.

December Tunnels completed.

1984

February	Reorganization of PSNH management.
March	Joint owners conditionally vote to cancel Unit 2 by December.
April	Temporary suspension of construction because of financial constraints; 5,200 workers at site laid off.
June	New Hampshire Yankee organized; new financing secured for PSNH; construction resumed at reduced pace.
November	Secretary of Energy Hodel pledges administration's support for completion of first reactor.

1985

December	NHRERP submitted to FEMA for review.

1986

February	NHRERP tested in formal exercise; FEMA concludes that the plans have major deficiencies. Nuclear fuel arrives on site.
April	Catastrophe at Chernobyl nuclear plant in USSR. Governor Dukakis orders halt to all emergency planning in Massachusetts.
June	New Hampshire submits revised emergency plans. NHY applies for low-power license.
July	NRC directs Shoreham licensing board to consider utility-devised emergency plans and to assume that governmental officials will carry them out.
September	Governor Dukakis announces that Massachusetts will not submit emergency plans for communities in the ten-mile radius because the public safety cannot be assured. The ASLB charged with on-site emergency planning opens hearings in Portsmouth.
November	Unit 2 canceled.
December	NHY asks NRC to reduce the emergency planning zone from ten miles to one. New Hampshire Supreme Court rules that local communities may not block or veto state efforts to develop emergency plans.

1987

January	NRC stays low-power testing permit until a decision about emergency planning during that phase of testing is reached.
February	NRC staff proposes rule change permitting substitution of utility-devised emergency plans when state and local governments refuse to participate; adopted in October.
April	NHY submits emergency plans for Massachusetts; NRC rules in June that the plans are not adequate. NRC decides to require emergency plans prior to issuing low-power license. ASLB denies NHY petition to shrink the emergency planning zone.
May	ALAB revises schedule for hearings on NHRERP, which will commence in September.
July	PSNH admits major financial difficulty in SEC filing; intervenors ask reopening of ruling on financial qualification; denied in August.
August	U.S. House of Representatives defeats measure that would have given state and local officials power to block nuclear plant licenses because of perceived emergency planning deficiencies.
September	NHY submits revised emergency plans for Massachusetts.
October	ASLB hearings on emergency planning open; Judge Ivan Smith replaces Judge Helen Hoyt as chairman of the panel.
November	FEMA testimony declares NHRERP inadequate; revised in March 1988.
December	NRC submits NHY emergency plans for Massachusetts to FEMA for review.

1988

January	New Hampshire Supreme Court rules prohibition of CWIP is constitutional. PSNH files for bankruptcy under protection of Chapter 11.
March	Massachusetts asks ALAB to stay low-power testing pending decision on PSNH financial qualification.
April	NRC staff proposes rule change eliminating need for off-site notification systems during low-power testing; adopted in September.

May   Hearings resume on emergency planning.

June   Massachusetts Municipal Wholesale Electric Company announces that it will stop its 12 percent operating support for Seabrook.

Two-day exercises of emergency plans for Massachusetts and New Hampshire called satisfactory; FEMA formally endorses plans in December.

September   Bankruptcy court denies motion to make NHY independent of PSNH.

Circuit Court upholds NRC rule allowing substitution of utility-sponsored plans for those of states.

NRC rules that low-power testing will be delayed until NHY provides assurance of adequate funds to assure safe decommissioning.

November   George Bush wins presidential election; names New Hampshire governor John Sununu chief of White House staff. Judd Gregg succeeds Sununu.

President Reagan orders FEMA to provide emergency plans when state and local governments refuse.

December   NRC requires bond of $72.1 million for decommissioning prior to low-power testing.

PSNH files bankruptcy reorganization plan, which establishes wholesale generating company under federal jurisdiction.

ASLB approves NHRERP.

1989

March   Governor Gregg announces intent to block low-power testing until agreement on rates is reached.

NHY submits bond to assure availability of funds for decommissioning.

Hearings on utility-sponsored emergency plans for Massachusetts commence.

April   Governor Gregg restates and then retracts his opposition to low-power testing after PSNH stipulates that potential cost of decommissioning will not be passed to ratepayers.

May   NRC authorizes beginning of low-power testing.

June   Low-power testing begins.

Seabrook Station

1 || Introduction

"What can you do with a dead dinosaur?"

The emotional thermometer at the Philadelphia headquarters of United Engineers and Constructors registered something close to angry early in 1977. Over several years, the firm had dedicated almost two and a half million skilled and costly hours to the design of the nuclear power station eleven New England utilities proposed to build at Seabrook, New Hampshire. The plant responded to a local need for cheap electricity and the national need for reliable energy not derived from petroleum, which in New England tended to be imported and expensive. Construction would provide a profitable market for dozens of industries and employment for thousands of people at a moment when the nation's economy could use a boost. But Lilliputian lawyers and bureaucrats, exploiting a perverse governmental process, were combining to withhold this Gulliver-sized bonanza from the public. The "saga," wrote the company's director of licensing to one of those governmental functionaries, was, for an "initiate," "a technical and economic nightmare"; for the "outsider," it was "an absolutely incomprehensible enigma." "We, who have spent our professional lives in the fathering and raising of the nuclear industry," H. L. Bermanis continued, "are beginning to despair," a reaction triggered in part by "the apparent indifference of the federal establishment."

Thomas Dahl, who was a couple of rungs up the corporate ladder from Bermanis, tried to enlist the governor of Pennsylvania in an effort to break through that perceived federal indifference. Perhaps Governor Milton Shapp, a Democrat, might be able to prod the administration of incoming Democratic President Jimmy Carter to develop a rational licensing policy for nuclear power plants. Dahl itemized Pennsylvania's stake in the Seabrook project: about $300 million in contracts, a thousand jobs, and a great deal more than that in the long run. And he charted too the stop-and-go regulatory process that was strangling the enterprise. The Nuclear Regulatory Commission (NRC) had wasted a year and a half approving the plant's design; the Environmental Protection Agency (EPA) was about to waste several more months with a second look at the "cooling water system," which was "about as good . . . as is reasonable and practical from an engineering standpoint" and environmentally superior to those in use at other coastal generating stations. Any proposed alternative was "simply not feasible."

To show that the company had friends in high governmental places, and perhaps to prove that his irritation was not unique, Bermanis enclosed Dahl's letter with his own to Roger Boyd at the Nuclear Regulatory Commission. The reply, which was tardy, contentless, and, in a sense, faceless, since it did not come from Boyd, unintentionally illustrated some of the points Bermanis had made. "Hopefully," it read, "the licensing process will become simpler in the future, especially in view of the attention it is currently receiving." There was no apology for the substance of policy, though perhaps greater simplicity in applying it was in order; nor did snarled construction calendars, disrupted careers, and bureaucratic delays themselves impart as much urgency as "the attention" they attracted. The reply was mailed several weeks after Bermanis's letter, a gap that furnished an inadvertent index of both the agency's responsiveness and its conception of punctuality.[1] And, of course, the delay of which Bermanis complained was only the beginning.

The Public Service Company of New Hampshire (PSNH) began to plan for a nuclear generating station in the 1960s, when atomic power had a bright commercial future, the nation's economy seemed stable and prosperous, the prestige of the scientific and engineering professions, enhanced by the exploration of space, was high and rising, and popular confidence in the government and other established authority had not yet been battered by the traumas of Watergate, Vietnam, and other events of the following decade. A relatively small electric utility, under conservative managers who had been with the company for years, PSNH served about 80 percent of a state that seemed poised for growth in both population and economic activity, and consequently for a surging demand for electricity. Late in the 1960s, together with United Illuminating, a Connecticut holding company, PSNH explored the possibility of building an 860-megawatt station at Seabrook. When United Illuminating backed out, PSNH, which could not finance the project alone, put the plans on the shelf.

They did not remain there long. Revived in 1971, the scheme attracted new investors and grew from one unit of 860 megawatts to two of 1,100 megawatts each. United Illuminating would own 20 percent of the facility; Northeast Utilities, another Connecticut holding company, 10 percent; and several other New England power companies would divide 20 percent. PSNH, which would manage the plant's construction and eventual operation, was to provide half the capital and receive half the output.

It was an ambitious undertaking for managers who knew about coal and oil but had no nuclear experience. With a projected share of almost $500 million, it was also an expensive enterprise for a company whose largest previous investment had been an oil-fired 476-megawatt plant that had cost about $80 million. PSNH's financial statements for 1972 showed a profit of about $2.25 million, of which more than $2 million was a noncash accounting entry. [2] The cost estimate, which was to prove wildly inaccurate, made no allowance for a national climate that had begun to shift since the project's conception some years before. No one in 1971 could anticipate the rapid increase in interest rates or the extent of inflation, pushed by rising energy prices, that would plague the coming decade. Executives ought to have noted eroding confidence in government and more generally in authority, but they probably did not see that the phenomenon applied in any way to Seabrook Station. They perhaps wrote off the new interest in the environment as a momentary fad or nuisance, and they surely underestimated the importance of the National Environmental Policy Act (NEPA), passed two years before. They knew in a general way the regulatory hazard they faced, but they expected the merit of the project to be evident quickly to sensible officials and the public, and they expected to prevail. Their expectations were not unreasonable, but they were disastrously wrong. Yet management clung to them with a stubborn tenacity worthy of flinty New Englanders of stereotype, in spite of costs that climbed more than ten times, deadlines missed by years, opposition that grew increasingly formidable, dwindling resources, and eventual corporate bankruptcy.

This reluctance to give in to change was not confined to New Hampshire's utility. The state proudly lived under laws passed by a part-time legislature and a Constitution adopted during the American Revolution. Revenues came from liquor stores and nuisance taxes; at the slightest hint of a tax on income or sales, voters rendered a political death sentence. A small state budget bought few state services: highway maintenance, underfunded state hospitals and colleges, dedicated state employees more distinguished for loyalty and reliability than for inspiration and ambition. Geographically compact and in 1970 still quite rural, New Hampshire nurtured a traditional suspicion of the less frugal, more urban ways of neighboring Massachusetts and lived on a small-town scale.

New Hampshire's small towns treasured independence. They raised their own taxes, paid for their own schools, fire trucks, and snowplows, and minded their own business. People read the state's major daily newspaper, the idiosyncratically conservative *Manchester Union-Leader,* and listened to Republican politicians, but they made their own decisions at annual town meetings. Authority in the community stemmed from ability and long acquaintance as much as from wealth or credentials that impressed a wider world. Citizens defended their own independence as well as that of the town, and they respected self-reliance in others. You stand on your own feet; you pay your own way. Live Free or Die, the state's motto, was descriptive as well as hortatory.

But New Hampshire did not escape the economic and social change and the cultural trauma that overtook the other forty-nine states in the 1970s. Old-time residents blamed transience and creeping urbanization on the invasion of Massachusetts. High-tech industry sprawled across the New Hampshire border from the corridor around Boston. The *Boston Globe* and Boston television stations reduced the influence of the *Union-Leader* and increased receptivity for Democratic politicians, a few of whom even won office. Rural concern about fish and game merged with growing environmental consciousness as housing developments replaced trees, streams, and open space in the southern part of the state. The web of relationships that held communities together became more tenuous as strangers moved in, regional schools and malls sprang up, and lodges, bowling leagues, and churches withered.

But if the society was in flux, New Hampshire's political institutions were not. Broad-based taxes remained a forbidden topic of political discourse. Both state and local governments spent money with great reluctance. Legislation to control growth or restrict land use sometimes passed, but traditional attitudes about property rights and individual freedom made enforcement difficult. A few more Democrats claimed places at the state Capitol in Concord, and a small volume of legislation reflected the state's rapid development, but political change lagged.

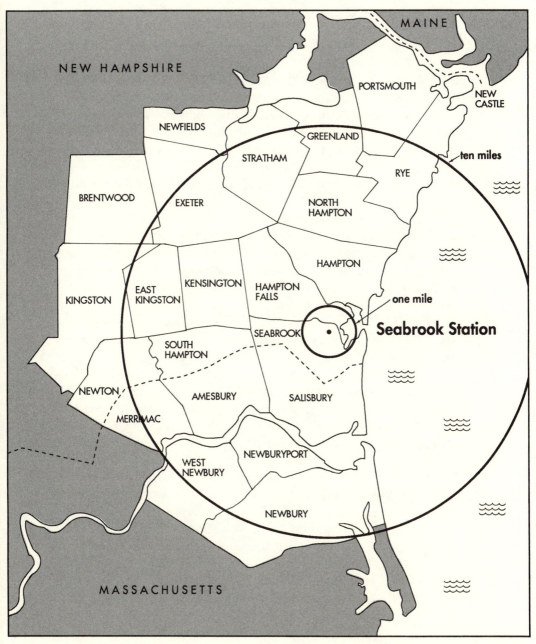

NEW HAMPSHIRE

MAINE

PORTSMOUTH

NEW CASTLE

NEWFIELDS

GREENLAND

ten miles

STRATHAM

RYE

BRENTWOOD

EXETER

NORTH HAMPTON

HAMPTON

KINGSTON

EAST KINGSTON

KENSINGTON

HAMPTON FALLS

one mile

• Seabrook Station

SEABROOK

SOUTH HAMPTON

NEWTON

AMESBURY

SALISBURY

MERRIMAC

NEWBURYPORT

WEST NEWBURY

NEWBURY

MASSACHUSETTS

Seabrook and surrounding communities

The state's largest utility liked it that way. One of the few enterprises in New Hampshire to employ a full-time lobbyist, PSNH had been able to do business for decades with relatively little inconvenience from state government. Guaranteed by law a return on investment, the utility usually received from the understaffed rate-setting Public Utilities Commission about what it requested, a situation that did little to promote aggressive management or corporate efficiency. Nor was there much need to ruffle the cozy relationship between state and corporation in 1970, when the legislature enacted a law, which PSNH did not oppose, establishing a new committee to approve the location of proposed generating stations. The fifteen members of the siting committee had other, full-time administrative duties, and the legislature provided neither staff nor detailed guidance about procedure. The new arrangement—another manifestation of government without expense—no doubt seemed to PSNH both harmless and tolerable, and so, after some months of desultory hearings, it turned out to be.

The delay those hearings caused could not have been avoided anyhow. While the siting committee pondered, the utility's engineers and legal counsel began a long dialogue with the staff of the Atomic Energy Commission (AEC). In effect, those conversations were a rehearsal for the federal licensing procedure that would follow. The staff reviewed the proposed plant's design, equipment, and location, and arrangements to protect employees, the environment, and the public. Any serious reservations had to be corrected before formal licensing commenced, a process that ordinarily committed the agency's staff to support the utility's application. In the case of Seabrook, this negotiation took place while the state was deciding to approve the site.[3]

Simultaneously, Congress decided that the twin responsibilities of the Atomic Energy Commission—to promote and to regulate civilian use of atomic power—were potentially in conflict. Just as the Seabrook application moved up the AEC's calendar, the regulatory functions were assigned to a new Nuclear Regulatory Commission. The new agency also inherited rules, procedures, and staff, and that continuity made the reorganization inconsequential for the project at Seabrook.

Congress vested the five commissioners of the Nuclear Regulatory Commission with executive, judicial, and legislative powers. The commissioners may issue rules governing the construction, licensing, and operation of nuclear power plants, decide conflicts over interpretation and application of those rules, dispense licenses to build and run a plant, inspect materials and workmanship, and permit or interrupt production of power. The agency has a large technical and legal staff to carry out its primary responsibility, which is to protect the health and safety of the public. The judicial function is delegated in the first instance to Atomic Safety and Licensing Boards (ASLB). These three-person panels typically consist of an administrative judge and two members with scientific expertise. Disputed rulings may be appealed to a similarly constituted Atomic Safety and Licensing Appeal Board (ALAB). The commission itself sometimes acts as a court of last resort, though any party to any case may also try to move the controversy into the federal courts.

A map of the NRC's judicial hierarchy only begins to describe the legal complexity the applicant for a license may encounter. A construction permit is the first step, which may be authorized by an Atomic Safety and Licensing Board after adjudicatory hearings. In the case of Seabrook Station, these first hearings dragged on for about a year and a half. Members of the public who wish to contest the application petition the ASLB to become intervenors and, if accepted, may explore an applicant's plans, cross-examine those responsible for them, and present witnesses with contrary views. Each side has access to the bases of the opposing case through discovery questions and answers; lawyers are supposed to anticipate their opponents' arguments and to disclose in advance the evidence on which their own client relies. The ASLB then shapes the debate and the proceeding by accepting, and sometimes amending, legal assertions, called *contentions*, sponsored by intervenors. The adjudicatory hearings on these contentions resemble a trial with slightly relaxed rules, at the conclusion of which the ASLB authorizes or denies a construction permit.

Thomas Dignan, the Boston attorney who represented PSNH from the outset, has maintained that adjudicatory hearings are the wrong way to settle licensing disputes. His criticism cannot be dismissed as sour grapes, for none of the dozens of lawyers in the years of Seabrook litigation has used the system more effectively or represented clients with greater skill. Nevertheless, Dignan has pointed out, the process can deal most satisfactorily with relatively simple questions, and those associated with licensing, because they depend on judgment and not on fact alone, are complex. Safety, for example, is only partly susceptible to a factual demonstration; the real question is not, Dignan has argued, whether a plant is absolutely safe but whether, considering the great benefit to be derived from its operation, it is safe enough, which is, of course, a matter for judgment.

Richard Meehan, another critic of the adjudicatory method of settling licensing disputes, has given a specific focus to Dignan's objection. An engineer and consultant to General Electric and other firms interested in nuclear licensing, Meehan has said that this form of inquiry is entirely ill suited to establishing scientific fact. The "proceedings are controlled by attorneys," Meehan notes, who "are . . . not interested in the scientific questions" or even "in the answers." Lawyers "are interested only in the written record," because "they know that it's going to be appealed and that what the written record looks like is what counts."

The licensing board that creates the written record, Meehan says, or at least the members with scientific training, might grasp "a clear, simple presentation" of scientific data, "perhaps at the freshman survey course level." But "it seems doubtful that the panel members follow the entire intricate course of argument" about complex scientific matters. The judges listen, Meehan says, but they cannot in most instances make an independent scientific judgment; rather, they decide which witness, which advocate, is more credible and then adopt that point of view. Lawyers with wealthy clients, aware of this tendency, simply "buy the biggest name," as one of them told Meehan, to carry the case. And technical bureaucrats, who are supposed to advise licensing boards, sometimes propose studies or other delaying devices in the hope that a consensus will emerge from legitimate differences of scientific opinion. [4]

Documents the NRC staff itself must prepare also may betray reluctance to take a position on disputed scientific issues. Both an environmental impact statement and a preliminary safety analysis must be in the record before hearings on the application for a construction permit can commence. Both require a quasi-scientific assessment of material supplied by the utility and its consultants and vendors; both are released in draft form for public comment and criticism. But comments on the environmental study prepared for Seabrook Station, for instance, were not adopted or refuted or used to modify the text; rather, they were simply printed as an adjunct to the prepared statement, a practice that just passed along disagreement, and perhaps confusion, to the ASLB for resolution.

Congress intended an independent Advisory Committee on Reactor Safeguards (ACRS) to provide authoritative, scientific advice and to serve as a check on the NRC's technical staff. Most members of this body are professors of academic disciplines relevant to the production of nuclear power. They are, in effect, part-time consultants who assemble for a few hours or a few days to review plans for power plants and to consider matters in need of further research. The ACRS has asked important questions about nuclear technology and public safety. Some members have charged that not every query has received careful attention or a satisfactory response. But the committee has usually kept its misgivings out of the public record and permitted licensing to continue.[5]

Although the Atomic Energy Act of 1954 has been amended, the regulatory jurisdiction of the AEC, and subsequently that of the NRC, has not been directly impaired. But other legislation has hedged that authority, and other federal agencies have developed a legitimate and sometimes a statutory interest in the expansion of nuclear power. The National Environmental Policy Act of 1969, for instance, required an analysis of the environmental consequences of federally sanctioned initiatives; subsequent amendments to the act gave the Environmental Protection Agency jurisdiction over water pollution. The judiciary eventually ordered a reluctant NRC to provide thorough studies of the effect of nuclear plants on the environment, and the EPA asserted a decisive role in evaluating cooling systems. The NRC did not suffer gladly either perceived infringement. And there were others—from the Federal Emergency Management Agency, with respect to emergency planning; the army's Corps of Engineers, with respect to rivers and harbors; the United States Geological Survey, with respect to past and potential earthquakes; the Federal Energy Regulatory Commission, which regulates wholesale electric rates; the Coast Guard, the Commission on Historic Preservation, and others that must be consulted.

Then, once a licensing board has held adjudicatory hearings and synthesized all this material in an initial decision that itemizes findings of fact and conclusions of law, those who disagree appeal. The Atomic Safety and Licensing Appeal Board will not necessarily rehear every disputed item, but the panel may make a much more wide-ranging inquiry than can an appellate court. The ALAB may review evidence as well as argument, fact as well as law. The Seabrook appeal board essentially rewrote an initial decision the judges concluded was defective. Those litigants who remain dissatisfied may renew their appeal to the NRC itself, to the courts, or in certain circumstances to the quasi-judicial processes of other federal agencies. Appeals, in short, may follow many different paths, separately or simultaneously, and go on literally for years.

To mitigate the extraordinary cost of delay, the NRC has permitted applicants to commence construction as soon as a permit is issued, even though appeals may be pending. Although the procedure has since been amended, the construction permit for Seabrook Station took immediate effect, and the legal burdens for those who sought to halt or interrupt work at the site were difficult to meet. Since an appeal board might have revoked the permit, investment involved some risk, a peril that was markedly reduced by the NRC's willingness to consider expenditure at the site in subsequent cost/benefit analyses. The "immediate effectiveness" rule, then, skewed the appeal process and left the judges of the First Circuit Court of Appeals somewhat bemused when they became involved in the Seabrook case. "We are," they wrote, "unable to identify any other field of publicly regulated private activity where momentous decisions to commit funds" rested on "preliminary decisions" that any of several governmental bodies might reverse. To be sure, the court continued, the risk to private investors was "real and always present." But "perhaps more important" was "the risk that public agencies and courts" would "accept less desirable and limited options, or worse," acquiesce in "a *fait accompli.*"[6]

As construction—and often the appeal process—nears completion, owners must apply for an operating license and thereby trigger a regulatory rerun. The notion behind this second stage is that every design problem need not be solved at the outset and that engineers ought to be free to take advantage of techniques and technologies developed while the plant is under construction; new regulations may also impose new requirements that must be met before operation. A final set of adjudicatory hearings provides one last check on the quality of equipment and construction and the qualifications of operators and thereby assures public health and safety. Before the accident at Three Mile Island in 1979, the NRC's director of reactor regulation testified, approval of an operating license was the nearly automatic result of a "stylized, ritualized process." The enterprise had been subject to the NRC's oversight from the beginning, after all, and the agency never found more than "minor adjustments" necessary at this late stage. Remaining unresolved safety issues were routinely postponed as "generic." The dollars and time invested in the project tended to encourage cursory review, often uncontested by weary intervenors, though a decision to authorize an operating license was, of course, subject to renewed appeal.[7]

No one admires this process, and in fact the NRC has simplified it since Seabrook Station received an operating license. The new procedure, which combines construction permit and operating license phases, ought to speed licensing and reduce costs and irritation, though it is not clear that public confidence in the outcome will be enhanced.

These amended rules respond more directly to suggestions from advocates of nuclear power than to charges by opponents that the entire regulatory process is unfair.

The contentious, costly, dilatory licensing of Seabrook Station was one of the object lessons that forced the NRC to modify the process. A flawed initial decision and appeals from virtually every party in the case; several excursions to the courts; a tentative approval of the cooling system by the EPA, followed by a reversal, an appeal, a court case, and approval once again; several profoundly inadequate investigations of alternate sites, which the National Environmental Policy Act required; two temporary suspensions of construction at the behest of one or another arm of the NRC—most of the flaws in the licensing process have been exemplified. "This case," the commission confessed in 1977, "has been widely depicted as a serious failure of governmental process." The issues had not been resolved "in a timely and coordinated way," and the result was "a system strangling itself and the economy in red tape."[8] Fifteen months later, a member of the commission implied an apology to PSNH, which had faithfully adhered to the NRC's instructions and was still stymied: "Seabrook," wrote Commissioner Richard Kennedy, "is the perfect example of everything that is wrong with the present licensing process." Unless the system were reformed, "the debacle," which the commission "helped perpetuate," might be "endless."[9] Governor Meldrim Thomson of New Hampshire, one of the most outspoken of Seabrook's advocates, spent his six years in office attacking with almost equal abandon environmentalists, civil disobedients, and the state's Fish and Game Department for blocking construction, and "a mad federal bureaucracy, a gutless national administration, and a stupid judiciary" for failing to push the project to completion.[10]

When proponents, politicians, and members of the NRC could not get the rules changed, intervenors and opponents stood little chance. Utilities could recover their litigation costs from ratepayers, and governmental agencies from taxpayers. But underfunded volunteer groups had no deep pockets to pay for lawyers, witnesses, research, and appeals. If construction continued while the bureaucracy churned, intervenors charged, neither the utility nor the regulatory agency had any incentive to reform the process; delay was almost as acceptable as approval and would cost intervenors more. "If the aim of this strategy of protraction is to end the intervenors' participation," prophesied counsel for the New England Coalition on Nuclear Pollution (NECNP) in 1981, "it will surely succeed."[11] The NECNP, in effect, stood proponents' argument on its head. They had pointed with dismay to the ruinous cost delay inflicted on the owners of Seabrook Station—$10 million every month, Meldrim Thomson once claimed. But given their comparative resources, the NECNP argued, costs to intervenors were even more debilitating.

Congressman Michael Harrington of Massachusetts, whose opposition to the Seabrook project was early, calm, and sensible, gave a different twist to the argument that governmental inefficiency weighed most heavily on opponents. Regulatory agencies "all too often . . . get bogged down in the technical" details of a project, Harrington said. That approach delighted "industry because it has the expertise and the funds to win on the technicalities." But preoccupation with smaller matters was a bureaucratic shell game that diverted attention from decisive issues of policy. Licensing forced opponents to concentrate on cooling systems and transmission lines and details and barred consideration of whether nuclear plants ought to be constructed at all, at Seabrook or anywhere else. That situation, Harrington held, "must be reversed before we find ourselves wedded to a technology whose total impact" had not been foreseen. "The place to reverse" this pattern of "decision-making is here," he told the ASLB in the fall of 1975, "and the time is now."[12]

More than a decade later, with a restrained eloquence that distinguished her from the boisterousness of demonstrators at another ASLB session on Seabrook, Mary Metcalf once more made Harrington's point: Complex division of the licensing process, she observed, sapped the public's confidence, shifted the inquiry's focus from protection of the public to subordinate matters, and took forever. A retired elementary schoolteacher from Durham, Metcalf had become a self-taught expert on New Hampshire's Public Utilities Commission. She bought stock in PSNH and tried to curb the company's enthusiasm for Seabrook Station through proposals at the annual stockholders' meeting. She was small, almost retiring in her manner, but there was nothing wrong with her analysis.

Mary Metcalf knew that the board she addressed on that October evening in 1986 could not reply to or act upon her remarks; she was almost as familiar with the NRC's procedures as she was with those of the Public Utilities Commission. Virtually every matter she discussed was outside the purview of the shell-shocked, weary judges who decided to allow the public to talk in the evening in return for preserving a measure of decorum during official daytime proceedings. She apologized for the jurisdictional irrelevance of her remarks, and she tried to explain why she and others had to speak anyhow: "Writing letters to faceless bureaucrats at vague addresses does little to calm anxiety, especially if there is no response or the response is evasive in character." She knew she would receive no answer that evening as well, but Mary Metcalf went ahead regardless.

> My concern [she began] is the extreme fragmentation of this regulatory process and what, in my opinion, is the faulted decision-making process that is the result.
>
> You were not assigned this week the task of evaluating the integrity of the construction of the Seabrook Station. . . . But it is clearly of great public concern.
>
> You were not assigned this week the task of evaluating evacuation plans. But they are clearly a matter of public concern.
>
> You were not assigned this week the task of resolving the several nuclear waste storage problems. But they are clearly a matter of intense public concern.
>
> You were not assigned this week the task to determine the extent of health risks to residents and workers. But they are clearly of public concern.
>
> You were not assigned this week to consider the excessive borrowing costs imposed on Public Service Company because of the risks perceived by the financial community to be associated with the Seabrook station project. However, the inclusion of those costs in future electric rates is clearly of concern to the public.

Mary Metcalf knew where the responsibilities she had ticked off did belong.

> The integrity of the overall construction is assigned to the Nuclear Regulatory Commission, which is divided—and I emphasize divided—among many branches, boards, and panels. You represent only one of these.
>
> But at the same time, some other construction issues—for example the cooling tunnels and the attendant thermal pollution—are assigned to the separate Environmental Protection Agency, which too is further divided and subdivided. Evacuation planning is assigned to yet a different federal body, the Federal Emergency Management Agency, which undoubtedly is subdivided into specialty areas. And the nuclear fuel waste problem is, in itself, divided, with the federal Department of Energy responsible for high-level waste, and with the State of New Hampshire responsible for low-level waste, and with greater-than-class-C waste still searching for a sponsor.
>
> Financial issues—both the borrowing of money to underwrite the con-

struction costs and the ultimate ratepayer issues—are determined not by a federal agency, but by a state commission, the New Hampshire Public Utilities Commission.

"This fragmentation of the regulatory process," Metcalf continued, had produced "chaos . . . confusion, . . . and the personal frustration that has been often expressed here these past several days."

> If the public had confidence in the integrity of each of these several regulatory agencies, there would be no problem. Lack of such confidence has been expressed repeatedly here this week.

> If the public had confidence in Public Service Company of New Hampshire, there would be no problem. Lack of such confidence has been expressed repeatedly this week.

Metcalf urged the judges to deserve her confidence and that of the rest of the audience. She also deserved an acknowledgment and perhaps their applause. But the ASLB, running late and understandably numb from the pummeling it had absorbed, resignedly called up the next speaker.[13] "When one looks at all the factors at work in the Seabrook regulatory soup," Donald Stever reflected some months after he had extricated himself from the kettle, "one begins to get an uncomfortable feeling that the truth . . . became secondary to a more compelling need—the termination of a process that had gone on too long."[14]

One unanticipated consequence of the lengthened process was an erosion of the credibility of those who supported Seabrook Station. Management, in particular, which made periodic public predictions about the rising need for electricity, the cost of construction, and the date of operation, looked increasingly foolish when consumption was steady, costs doubled and then doubled again, and deadline after deadline passed. Granted the notoriously dilatory licensing process explained, and to some extent excused, those embarrassingly visible errors; if the expected calendar had been met, both the scope and the embarrassment of the mistakes would have been reduced, though errors (and the bills for them) would have remained. But a large and unlicensed nuclear plant was a constant and highly visible billboard bearing a message that managers had miscalculated matters about which they were presumed to know. And if they could not correctly predict costs and demand, detractors wondered, was it likely that the same fallible people could safely manage a technologically complex facility, a task for which they had no relevant experience?

Delay also exposed a growing number of promises the nuclear industry had failed to keep. Nuclear electricity was not in fact too cheap to meter, as its prophets had once predicted. Nuclear reactors did not just hum along without cost or interruption; some were shut down as much as a third of the time. *The China Syndrome*—a movie about a core meltdown—was not a Hollywood fantasy but might, it seemed after Three Mile Island, be playing at a nuclear power station near you. Accidents that engineers defined as "incredible" occurred not only at Three Mile Island but, less notoriously, near Detroit. The assurance that radioactive waste could safely be stored in a breakfast nook seemed not only mistaken but disingenuous when, after decades of searching, the federal government still had no storage facility. The major nuclear explosion that could not happen had obviously occurred in the Soviet Union. And so on.

Had PSNH been able to keep government, contractors, designers, labor unions, banks, the stock market, intervenors, demonstrators, and other interested, but dawdling, parties on the schedule management proposed, Seabrook Station might not have become the warehouse for the industry's inventory of unkept pledges. Seabrook was supposed to have been completed about the time of the accident at Three Mile Island in 1979. The federal government seemed then to have a reasonable, although untried, scheme for disposing of nuclear waste. The profound economic and psychological shock of the oil boycott and the escalation in energy prices was then fresh in the popular mind and would have numbed somewhat outrage about the price of nuclear electricity. Chernobyl would not have been part of any American's vocabulary.

But, of course, Seabrook Station did not open on PSNH's calendar in 1979—or for a decade thereafter. And, although that licensing delay probably magnified some of management's miscalculations, the lapse of time exposed, more than it created, errors. That exposure occurred, furthermore, when a suspicious national audience had been conditioned to mistrust authority, to second-guess experts and denigrate their expertise, and to assume that relevant information had been concealed from the public. Predictions of painless victory in Vietnam rang in the nation's memory, as did the protestations of innocence of a disgraced president. The marvels of the space program had not yet disintegrated in the agonizing moments of the *Challenger* explosion, but the nation's engineers clearly found garbage an intractable problem and could not produce a car that did not have to be recalled. American spies seemed

inept, and America's police could not locate any of a half-dozen notorious fugitives. Inflation plagued a stagnant economy, a condition that
baffled economists, and the interest rate defied imagination. The later
1970s, in short, were not a period when the people of the United States
expected routine scientific miracles, automatically increasing prosper-
ity, and the whole truth from people in authority. That was not an
atmosphere conducive to the construction of nuclear power plants, at
Seabrook, New Hampshire, or elsewhere.[15]

"[T]here was a time," Richard Meehan recollected, "when experts
exercised special powers, when a man could present certain qualifications to the world and be believed thenceforth without serious question." His experience with nuclear licensing in the 1970s had persuaded
him that "the powers of expertise are experiencing a fate similar to the
dollar's." In fact, Meehan thought, the declining authority of expertise
was partly the fault of the dollar; certified advice on any side of almost
any issue was for sale. In his own nuclear consulting, Meehan had
encountered cynical utility executives who made decisions and then
bought experts to justify them. The trouble with the game was that
others hired contrary authorities, and (putting the situation in the most
favorable light) a legitimate difference of scientific opinion was passed
along to a licensing board, which had the authority and the duty, but not
necessarily enough knowledge, to discern the truth. Utility commissions
in New York and Wisconsin, for instance, after examining reams of
contradictory testimony, could not conclude whether the output of a
nuclear station would cost more or less than that derived from fossil fuel.
Scientists and engineers, even if uninfluenced by the source of their
fees, may have differing presuppositions that determine their interpretations of data. What looks to a geologist like a terrible risk, Meehan says,
seems an interesting challenge to an engineer who wants to "take the
problem into account" and render a potential earthquake harmless
through design. Thus, a geologist's opinion about nuclear safety and, by
extension, opinions of other scientific authorities may, he adds, "be no
less ideological than the views of an Islamic mullah on women's libera-
tion."[16] Experts on seismology who advised Seabrook licensing boards
included a reputable consulting firm, a professor with a new and controversial methodology, and a housewife from Hampton, who more than
held her own. The issue traversed the NRC's regulatory hierarchy for
years until exhaustion, rather than reason, closed the controversy.

Experts who provoke skepticism and contrary opinion may invoke
reputation and credentials to buttress an opinion and may reveal a

patronizing impatience toward those who challenge them. The staff of the Nuclear Regulatory Commission has not always repressed displays of this sort of professional arrogance, nor, in spite of a belated conversion to skillful public relations, has the Public Service Company of New Hampshire. To a customer who wondered in 1977 whether the corporation was really wise to be so thoroughly committed to Seabrook, a frustrated corporate image polisher exploded that her attitude was un-American; he said, the customer reported indignantly, "that if I didn't like things as they are here, why didn't I go live in Russia."[17] In addition to irritation and an obvious lack of tact, the response shows the condescension of professionals who resent a cultural climate and, in the case of PSNH, a regulatory situation that confer on amateurs the status of experts.

By its very nature, therefore, the whole process of public participation in nuclear licensing can be seen as cynical theater. The inquiry is not open and never disinterested. Experts have already reviewed and approved the plans of successful corporate executives, who want to invest a great deal of money. Opponents are naive, do not understand, have bad information and insufficient training, and ought to calm down, listen to those who know, obey higher authority, and buy their electricity from Seabrook Station. The picture is overdrawn, but nuclear opponents, who have repeatedly encountered the condescension of the nuclear establishment, do not think it caricature.

Johnny Carson provided that on the "Tonight" show in May 1977. "Put me down," Carson joked, "as one American who favors building nuclear plants." Indeed, there ought to be "one on every street corner next to a McDonald's."

> I say we should trust science. Remember science has given us cyclamates, saccharine, and DDT. . . . DDT has worked so well we don't have to use it any more because it's working everywhere, in the rivers, in our food, and in our lungs. And what's all this fuss about plutonium? . . .
>
> They say if there is a leak in a nuclear power plant the radiation can kill you. . . . They say atomic radiation can hurt your reproductive organs. My answer is so can a hockey stick. But we don't stop building them.[18]

Nor, in spite of Johnny Carson, did they stop building at Seabrook.

Nuclear regulation depends not only on law and on the authority of scientific knowledge but also, in a political democracy, on the informed consent of interested citizens, which is presumably the purpose of public participation in the nuclear licensing process. That those extended proceedings, affecting Seabrook and other projects, were not exactly serving that purpose was no secret. Yet nuclear power—cheap, non-polluting, potentially inexhaustible—still appealed to presidents and less eminent politicians seeking an energy panacea. All that was required, they thought, was an accelerated bureaucratic procedure that would disseminate the convincing evidence and license plants more quickly. By 1989, President Richard Nixon predicted, nuclear fuel would furnish perhaps 40 percent of the country's electricity.[19]

That prophecy looked foolish before Nixon was well settled in private life. Between 1975, when the NRC opened hearings on the construction permit for Seabrook, and the accident at Three Mile Island in 1979, orders for more than twenty reactors were canceled, and more than forty other projects were delayed. By 1985 plans had been scrapped for an additional seventy-five plants, more than a third of which were under construction when abandoned. Since 1978 no utility has ordered a new plant.[20]

Explanations for this astonishing collapse are nearly as numerous as the cancelations. Most of those explanations, however, ultimately rest on somebody's judgment that the project was going to cost too much. One important component of that cost was the start-to-finish delay that rendered a very large asset unproductive for years. Opponents of nuclear power, whose political sophistication had been seriously underestimated by regulators and utilities, did delay completion or, in a few cases, block the operation of a generating station. But it was not the ragtag demonstrators, the environmentalists, or those with acute radiation anxiety who shut down the industry; it was the industry's customers, who concluded for various reasons that nuclear fuel was an uneconomical way to produce electricity.

Delay was a convenient scapegoat as well as a partial reason for rising costs. Neither bureaucracy nor inefficiency has a large or vocal constituency; both are anonymous and impersonal. Utilities, dependent upon regulatory approval, did not have to castigate any individual who might someday have to pass on a petition, though occasionally they did blame particular intervenors for intentionally slowing the process. Many politicians, including some involved in the Seabrook case, have made careers berating the bureaucracy, as if political leaders were powerless to counteract it by firing people, passing legislation, or providing effective executive direction. Delay, in short, was a convenient explanation that absolved utility executives, who had reflexively undertaken projects beyond the engineering and financial capability of their organizations, and other nuclear advocates, who had oversold the technology to the public. (Delay was also useful to those who actually had to design and build plants and who needed time to work out details that were beneath the elites who made policies and signed contracts.)

Bashing the federal bureaucracy appealed especially to New Hampshire politicians, who could rely on a sympathetic echo from fiercely independent voters, proud of the state's tradition of frugal, limited government. Governor Meldrim Thomson, for whom no problem had two sides and who relished his bull-in-a-china-shop image, personalized impediments to Seabrook Station, as did his editorial mentor, William Loeb of the *Manchester Union-Leader*. Thomson sacked state employees who wondered aloud about the benefits of the project, arrested demonstrators at Seabrook's gates, and intimidated those who sympathized with them. He used his leverage in the New Hampshire presidential primary of 1976 to remind President Gerald Ford of an unfulfilled promise to expedite Seabrook's license. Whether or not there was such a promise, the White House did intervene late in the fall campaign with a telephone call to the regional EPA official who had turned down the cooling system designed for Seabrook Station. John McGlennon stuck to his decision, which required two reversals by the EPA in Washington, a court case, and several months to undo. "I just felt the licensing of Seabrook Station became . . . inevitable," McGlennon said in 1986, "sort of like a locomotive, and there was nothing you could do to stop it." If he saw McGlennon's comment, former Governor Thomson must have thought the train almost a decade overdue.[21]

While Thomson flailed the EPA, Congressman James Cleveland, who headed Ford's primary campaign in New Hampshire, tried to hurry the NRC. In June 1975, Cleveland urged Chairman William Anders to keep the Seabrook hearings, which had just commenced, moving ahead. As a matter of policy, commissioners do not answer correspondence about specific cases, a punctiliousness that did not impress Congressman Cleveland when he received a response from one of Anders's subordinates. He had not tried to influence the commission's decision, Cleveland protested; he was simply noting, as had the president on many occasions, "that the very serious energy problem facing the nation calls for the very action I have requested—namely to move the damn decision along!" It was, Cleveland concluded, "the delay that bothers me and should bother you."[22]

Delay bothered everybody, and virtually everybody has studied it.

Study, of course, is a familiar device to avoid doing something, especially when multiple investigations produce results that conflict with one another and do not coincide with preconceptions. The nuclear industry and its political supporters expected to discover what a commissioner once facetiously called "the vampire intervenor" at the root of delay.[23] But the Federal Power Commission and the Library of Congress and the Department of Energy tended to absolve both intervenors and federal bureaucrats. "Events or decisions in the private sector unrelated to the regulatory decisions of the NRC," the Department of Energy concluded, accounted for "about 80 percent" of the delay reported in nuclear construction. Managerial reconsideration of consumer demand and cost, technical problems with design and construction, labor disputes, and bad weather, rather than bureaucrats, courts, and environmentalists, were holding up nuclear progress.[24]

Irresolution of management, to which the Department of Energy attributed most of the time lost on nuclear plants, was not descriptive of the situation at Seabrook before the financial crisis of 1984. As construction came to a halt there that spring, Congressman Ed Markey, whose Massachusetts district included towns abutting Seabrook, asked the NRC why the project had taken so long. Two months later—a lapse that was ironically apt under the circumstances—Commissioner Nunzio Palladino sent the congressman explanations furnished by PSNH, which Palladino thought reasonable but for which he took no responsibility. As of March 1984, when construction was suspended, PSNH identified fifty-three months of delay, of which stop-and-go decision making at the EPA accounted for thirteen. "Schedule reevaluation," "added labor requirements," "productivity losses," "severe condition of capital markets," and more "time than expected to meet some regulatory requirements" explained the remaining forty months. Palladino did not need to underline for Markey the fact that the company's own list blamed management and exculpated the NRC and intervenors.[25]

Blaming management also satisfied politicians, who might otherwise have had to modify or circumvent administrative procedures through legislation or influence but who would not, of course, invade the domain of free enterprise management. All the studies of delay did not themselves quiet criticism by Governor Thomson or President Ford or Congressman Markey, nor did they prevent future inquiries from other governors, presidents, and legislators whose intent was to cut red tape. The studies occasioned no discernible reform and no identifiable administrative action, and they could, therefore, be disparaged as contributing to the problem rather than providing a solution. "We have found," the president's commission reported after Three Mile Island, "repeated in-depth studies and criticisms both from within" the NRC "and without, but we have found very little evidence that these studies have resulted in significant improvement." The commission's indictment extended to topics other than administrative delay, but that was a case in point.[26] And the implication was that all the studies and suggestions, critiques and courts, political intervention and public disenchantment availed nothing; the process had a pace and a perverse momentum all its own.

In large part, that inertia flowed from the decision, taken in the Atomic Energy Act of 1954, to encourage private development of nuclear power. Circumstances since, including the revitalized concern for the environment and accidents at Three Mile Island and Chernobyl, have obscured, but not altered, that national objective, and the NRC has never lost sight of it. The agency's central function is the licensing of generating stations. They are supposed to do minimal damage to the surrounding region, and they are supposed to be safe. But they are supposed eventually to produce electricity. Before the application is formally received and before the licensing clock begins to run, the agency's staff has approved the project in principle. The site and general design have been reviewed; a preliminary safety report is in preparation; the environmental impact has been considered; personal contacts with management have ripened. Intervenors will subsequently be heard, and their objections, if well taken, may require altered design, an amended rule, and delay; if not serious, they will be ignored. The course of the process is not entirely predictable, but the outcome is.

Two events almost a decade apart seemed to Seabrook intervenors to prefigure their eventual defeat. Both instances involved the chief judge of the Atomic Safety and Licensing Board. In 1975 intervenors were cautiously optimistic that they were beginning to persuade Daniel Head that construction of Seabrook Station was not environmentally responsible; he was abruptly replaced. In 1983 intervenors accused Helen Hoyt of an unjudicial bias and sought her removal; she remained very much in the case.

Neither incident was quite so straightforward as partisans professed to believe. Daniel Head was indeed attentive to the arguments of intervenors and might ultimately have joined a dissenting colleague to form a majority denying the construction permit, though no one could be positive of that outcome. Although they had no direct evidence of conspiracy, Seabrook's opponents charged that Thomas Dignan, worried about Head's apparent objectivity, invoked the political assistance of Governor Thomson and, through him, of officials in Washington, who in turn made Head an offer he could not refuse. He resigned from the ASLB, accepted a federal post that never materialized, and rejoined the NRC, but not the Seabrook board, some months later. No intervenor believed that John Frysiak, Head's replacement, had an open mind.

Robert Backus, attorney for two environmental groups, immediately moved for "the administrative equivalent of a mistrial." The matter was of such grave importance, Backus said, that he would withdraw from the case unless the board were reconstituted. Frysiak was not stirred and "in the interests of justice" denied Backus's motion. The judge hoped Backus and his clients would remain in the case, but he would not inhibit their withdrawal if that were their wish. Backus absented himself briefly, a gesture that must have seemed even to him surpassingly empty, for he soon returned and has remained in the case ever since. Frysiak was responsible for the decision that permitted construction to start, a document that has been almost universally described as incompetent.[27]

Helen Hoyt, in charge of the board assigned PSNH's application for an operating license, radiated crisp determination in nearly every order she wrote. She did not always meet her own deadlines, but everyone else in the case would adhere to her brisk schedule or risk dismissal. No excuse—not the failure of the Postal Service or snowstorms or holidays or clerical errors at the NRC or the burden of unreasonable discovery questions—would deter Judge Hoyt from the timely completion of her task, which intervenors believed was the approval of whatever PSNH asked. She reproved with equal vehemence what she thought was tardiness on the part of individuals or states: "The orders of the Board are not invitations to be accepted," Hoyt told Jo Ann Shotwell, an assistant attorney general of Massachusetts, refusing to consider the Commonwealth's late-filed motion; "they are orders and will be acted upon as such."[28]

An evident personal antipathy between Hoyt and Shotwell flared during hearings in Dover in the summer of 1983. Following a sharp on-the-record exchange, Hoyt dismissed Shotwell and later reproved all the attorneys in the case for their lack of decorum, punctuality, and civility, which she would demand even if her insistence resulted in reversal on appeal. Shotwell replied that Hoyt herself was insulting: In "my entire life . . . in courtrooms and outside," Shotwell complained, "I have never been addressed by anyone in the tone in which I believe you address me repeatedly."[29]

A case could be made that Hoyt had not singled out Shotwell. The judge also excluded Guy Chichester, an antinuclear activist whom Rye, a small town north of Seabrook, had designated to observe the hearings. Chichester made remarks that Hoyt found disruptive and joined observers from other towns in an accusation that Dignan and an attorney for the NRC staff had coached witnesses during their cross-examination, a charge that enraged Hoyt. Apparently at her request, an initiative intervenors believed improper, Rye formally disavowed Chichester, and Hoyt barred his attendance. In turn, by mail, Chichester proposed that Hoyt disqualify herself, a suggestion she ignored. Her "injudicious support" for the applicants, Chichester wrote, and her "insulting" and "stifling" attitude toward opponents produced a "poisonous atmosphere" from which justice could not possibly emerge.[30]

Robert Backus, who had had his share of disagreements with Judge Hoyt, endorsed Chichester's request with a formal legal motion that was later echoed by the NECNP and Massachusetts. Backus knew he could not cite the scowls, glares, and body language that courtroom observers had interpreted as prejudicial; he enclosed newspaper clippings reporting a growing public suspicion that the outcome in Hoyt's court was foreordained. But, Backus said, the judge's on-the-record conduct—chastising counsel, belittling town representatives, arbitrary rulings, and control of the record—showed bias that deprived intervenors of their rights and constituted ground for removal.[31]

Thomas Dignan gallantly rose to Judge Hoyt's defense with a sharp attack on Seabrook opponents. Hoyt's rulings had been fair and proper; intervenors "brought to the proceedings a tension of defiance, obstinance, and provocation" that was "incessant and pervading." He suggested that Shotwell had deliberately sought expulsion, that intervenors played to the press because they could not persuade judges, a tactic that complicated Hoyt's task and amounted to "judge-baiting," and that Backus's motion was a last-ditch effort to delay the proceedings and create "administrative chaos."[32] Hoyt routinely denied Backus's motion, which bounced around in the NRC's appellate hierarchy for some months before the commission itself, with an admonition to "all concerned to adopt a more temperate approach," left Helen Hoyt in charge.[33]

The argument between Dignan and Backus, two able lawyers who had been involved in the Seabrook case from the outset, revealed much about the motives of both sides. In effect, Dignan accused Backus and other intervenors of bad faith. They were, he charged, deliberately creating controversy in order to appeal and delay the decision they could not legally prevent otherwise. The particular motion, Dignan implied, was not significant in itself but was simply part of a larger strategy to oppose nuclear power. Even in 1983 Dignan's perspective was lengthy; he realized that triumphs were rarely decisive, that only intermediate hurdles were cleared, that the concern of opponents often shifted from a search for a better cooling system or more effective emergency plans to the abandonment of Seabrook Station and then an end to the industry. Seabrook was a hostage, the president of PSNH had observed in 1977, for the commercial development of nuclear power: "If Seabrook is cancelled," as one of the company's executives said, "there probably won't be any more nuclear plants built in the United States for some period of time."[34]

That outcome would of course have been acceptable to Backus and the hundreds of people who helped pay his fees. But targeting Helen Hoyt was for Robert Backus not simply a tactical trick. She served for intervenors the same symbolic purpose Guy Chichester and other anti-nuclear ideologues did for the nuclear establishment. If they abused legal process to obstruct completion of Seabrook Station, she abused legal process to finish it; if they opposed nuclear power as a matter of principle, she favored it; if they were not scrupulous about means to the end, nor was she. And intervenors knew, as did Dignan and his clients, the bittersweet disappointment of inconclusive intermediate victories and of apparently heartening rulings that had no long-term significance because the Helen Hoyts of the world wrote the rules and then refereed the contest.[35]

The motives of Hoyt's critics, and those of intervenors in the Sea-brook dispute, were more complex than Dignan's brief conceded. Robert Backus had originally entered the case, for example, as the attorney for two environmental groups that dropped out when that phase of the controversy concluded. He was critical of the lunatic fringe of the antinuclear movement, whose growth he attributed to the "perceived unfairness of the NRC's licensing hearings" and from which he carefully differentiated the Seacoast Anti-Pollution League (SAPL), his client of longest standing.[36] That organization had evolved from a group of citizens whose first concern was protection of New Hampshire's coastal environment to one whose central purpose was prevention of the operation of Seabrook Station. Because of a growing reputation for effective advocacy, Backus also represented some of the New Hampshire municipalities that opposed the state's emergency plans. Neither Backus nor those he served were ideologically rigid foes of nuclear power; but, as Dignan correctly charged, they were quite happy to beat the plant with any club the licensing process offered, citing at various times safety, cost, environmental protection, property interests, and preservation of the charm and independence of traditional small-town life.

The circle of Seabrook's opponents always included groups beyond the immediate neighborhood. Some, like NECNP, were responsible adversaries of nuclear power in any location. Others, like the Commonwealth of Massachusetts, discerned public safety and emergency plans at the very beginning as the project's disabling flaw. The New Hampshire Audubon Society and the Society for the Protection of New Hampshire Forests shared SAPL's initial ecological concern. Accidents at Three Mile Island and Chernobyl, the federal government's notorious inability to dispose of nuclear waste, and fearsome projections of the future cost of power added other, later converts. Finally, in most cases because of the preparation of emergency plans, political entities assumed an increasing prominence.

Helen Hoyt tried to keep every intervenor on an equally short leash. She reproved impartially New Hampshire's assistant attorney general and ordinary folks who appeared in her courtroom. She threw out contentions sponsored by the Commonwealth of Massachusetts, officials of towns in two states, and attorneys such as Robert Backus and Diane Curran, whose participation in the case antedated her own. She routinely denied requests for additional time to prepare testimony or cross-examination, to reply to discovery requests, or to evaluate complex engineering or legal documents.[37]

But this one-dimensional impartiality did not, of course, signify that either the judge or the process over which she presided dealt equally with all litigants. Although she occasionally ruled against Dignan and even rebuked the NRC staff, both won more often than they lost. Nor did intervenors have equal access to the legal and financial resources that sustained the utilities and the NRC itself, though the active involvement of local governments evened the odds somewhat. Still, towns were often represented by amateurs and volunteers from whom Judge Hoyt expected the same efficiency she demanded of experienced counsel. She interpreted delay, on whatever pretext, as a personal affront, and she had ample opportunity to take offense.

Perhaps deliberately, Dignan served lengthy, detailed discovery questions on intervenors just before deadlines, a tactic that worked considerable hardship on individuals or part-time local officials who were supposed to reply completely and promptly. One exasperated selectman
loaded his car with cartons of papers that had been sent to the town and
dumped them, unacknowledged and probably unread, at the gates of the
plant. But refusing to answer Dignan's hundred-page questionnaires
could be cause for dismissal from the proceedings, as several towns
found out in 1986.[38]

The Hampton Beach Chamber of Commerce had occasioned a
similar ruling three years before. Beverly Hollingworth, a state representative from Hampton, owner of a motel at Hampton Beach, and an
energetic foe of Seabrook Station, spoke for the Chamber as well as she
could. She secured intervenor status, submitted contentions, and participated in the discovery process. But she was not a member of the bar,
she had to earn a living, and she could not devote every minute to her
unpaid effort to keep Seabrook Station inoperative. She was also sloppy
about deadlines, and her insouciant answers to Dignan's interrogatories
and those from the NRC staff amounted to what one lawyer called "no-
answer answers"; "I have been at this for seven or eight years," he
complained, and this was the first occasion when he had been "required
to put experts on a witness stand and make them available for lay cross-
examination" without advance disclosure of the examiner's interest.
Two weeks later, Hoyt dismissed the Chamber from the case.

Neither the incident nor the ruling had great significance. Hollingworth continued to assail the plant publicly and politically, a forum that
was beyond the reach of Hoyt's rulings. Counsel for other intervenors
cross-examined witnesses as skillfully as Hollingworth could have done;
she probably would have prolonged the litigation without making a
unique contribution. And, in fact, she had not lived up to her obligation
in the discovery process.

But her dismissal underlined a fundamental unfairness of what intervenors thought of as a one-sided system. I thought, Hollingworth told
Hoyt once, that "the American taxpayer whose interest was the safety
and protection of the American people" was the NRC's client. Yet the
agency provided her no legal assistance, expressed impatience with her
ineptitude, moved to disqualify her, and seemed to help make the case
for the utility, which already had high-priced lawyers of its own. Even if
Hoyt impartially required everyone to meet the same deadlines and play
by the same rules, that situation seemed to Hollingworth fundamentally

inequitable.[39] The NRC's own review of procedures after Three Mile Island concluded that, "insofar as the licensing process is supposed to provide a publicly accessible forum for the resolution of . . . safety issues . . . , it is a sham."[40]

The licensing system and the Helen Hoyts who try to run it have been the target of convenience for those dissatisfied with the glacial development of the nuclear industry as well as for those who believe any development is too fast. The process is vulnerable, and those who run it equally so. But in ridiculing rigged procedures, biased judges, incompetent staff, and sleazy commissioners, critics, including members of Congress, have waltzed around the central point. Personifying nuclear power tends to score political points while obscuring policy, allows harsh words to masquerade as action, creates no change, requires no understanding of complex technology, and costs no money. On both ends of the nuclear political spectrum, the temptation to attack opponents rather than advance solutions to the industry's intractable problems has become almost irresistible.

In 1987 and 1988, for instance, while $10 billion of invested capital sat idle at Seabrook and Shoreham, while the reactor at Three Mile Island remained contaminated nearly a decade after the accident, while generating stations owned by the Tennessee Valley Authority grew increasingly unreliable and perhaps unsafe, while the Price–Anderson Act, which provides insurance for the nuclear industry because private companies will not, was about to expire, and while radioactive waste piled up at generating stations across the land, Congress appeared to concentrate on the alleged unethical, and perhaps illegal, conduct of Commissioner Thomas R. Roberts. Roberts may have disclosed documents to a southern utility about an investigation in progress at an incomplete plant near New Orleans. His conduct, however interpreted, was one more link tying the NRC to the nuclear industry, a cooperative relationship that was open to abuse and criticism, but was hardly news.[41]

The incident was momentarily embarrassing, but the headlines disappeared and Commissioner Roberts remained at his desk. When motivated legislators cannot bring one ethically insensitive commissioner to bay, the prospect for a legislatively devised method of disposing of nuclear trash is not bright. Since a nuclear dump is even less of a neighborhood asset than a nuclear power plant, no one seems anxious to secure one for his constituency. And the projected cost merely to begin to control the problem created by military atomic waste staggers congressional imagination. In the summer of 1988, the Department of Energy estimated that $110 billion might be enough to stop the spread of radioactive material; the General Accounting Office thought the figure about $65 billion too low. Those figures included no allowance for actually disposing of the collected waste or for dismantling contaminated plants. Probably that was a prudent omission, since there was no place to put the toxic material anyhow. The Department of Energy expected to spend about $1 billion that year on environmental health and safety, an amount that seemed unlikely to make a major dent in the problem. Having discovered it, Congress backed away; "the costs appear to be so huge," admitted Senator William Roth, "that nobody wants to tackle it."[42] And nobody wanted to talk about one of the smaller ironies the hearings disclosed: Nuclear plants, on which billions were spent to make them indestructible, had to be literally disposable like so much American merchandise. At the end of a relatively brief productive life, they had to be "decommissioned," or somehow thrown away without so much as a radioactive trace.

Nobody, in fact, wanted to tackle commercial nuclear policy in any form. Unfinished or unlicensed nuclear stations had become visual metaphors for failed initiatives, bureaucratic constipation, inept management, and scientific fallibility. An incomplete plant manifested in concrete and rusting steel the gaps in the nation's energy policy, the hundreds of items of unfinished business on the agenda of Congress and of the regulatory agencies, the imperfect and extraordinarily expensive machinery that engineers had built and that management was attempting to sell. *Forbes* magazine used a cooling tower to introduce an article about "the largest managerial disaster in business history."[43] John McGlennon, the EPA administrator who disapproved Seabrook's cooling system, found another symbol to represent the persistence of the industry in spite of its manifold failings. Almost a decade after his decision set off a political and legal scramble that resulted in his pursuit of a career outside government, McGlennon advanced the "dinosaur theory" to explain Seabrook Station: "If you have a dinosaur in your back yard, you have to keep feeding it, because what can you do with a dead dinosaur?" Or, Seabrook's opponents wondered, with a live one, for that matter?[44]

2 The Environment

"Brick by brick, pipe by pipe, fish by fish."

Late in 1976 Thomas Dignan had heard enough about the natural beauty of the region where his clients wanted to locate their new nuclear generating station. For three years, he had listened to those who opposed construction describe the visual, recreational, and ecological importance of the site at Seabrook. Dignan knew about the warm sand and the cold water at nearby beaches and about the clams and lobsters that shared the shore and the surf with sun-seeking New England and Canadian families. He understood that "party boats" bore tourists seaward from Hampton Harbor to fish at ledges offshore and that lobstermen and recreational sailors used the protected anchorage as well; a few hardy souls lived for much of the year on boats tied up there. And Dignan also realized that the salt marsh that surrounded the harbor was no ordinary swamp; even the state of New Hampshire, which ordinarily granted the requests of his client, the Public Service Company of New Hampshire (PSNH), had so discouraged a scheme to bury piping in the wetland that the company decided, at considerable expense, to change plans. But the bemused lawyer wondered if those who praised the environmental wonders of Seabrook were in fact looking at the right real estate.

Before construction began earlier in the year, Dignan wrote in one of the endless stream of communications he sent federal agencies, "the most readily observed feature of the site . . . was the town dump." "The marsh," he continued, "is beautiful, but hardly pristine. A railroad runs through it, Route 1 runs through it, a transmission line runs through it, and some hundreds of miles of ditches have been dug, and remain, in it." Construction, Dignan assured the Nuclear Regulatory Commission's appeal board, would harm the "fragile estuary" less than any of the famous New England northeast storms that slammed the area several times each year. And the beach, he said, ought not to be confused with "Walden Pond"; rather, "it is an amusement area," complete with carnival rides, a dance hall, and an assortment of fast-food stands. "We bring this out," Dignan concluded, "not to denigrate the area, but to point out that Seabrook is not sited in an unspoiled glade previously populated only by nymphs and leprechauns."[1]

The derision was gentle, but the difference of perspective—and the consequent bewilderment on both sides—was entirely genuine. Those who advocated construction of Seabrook Station sincerely believed the proposed use of land and water would enrich the lives of tens of thousands of people. Opponents, with equal sincerity, found nothing progressive about nuclear fission and noted that the calculation of human enhancement, even if correct, ignored the catastrophic effect on millions of other forms of life—from cedar trees to plankton. The gap between these views, which often seemed a gulf as the controversy lingered through two decades, was not simply a matter of using different words to describe the same landscape. The semantic difference stemmed instead from fundamentally different views of the nation's past and future, from different sets of "facts," from differing estimates of risks and benefits, from, in the last analysis, differing values that did not blur or blend in the protracted regulatory process.

The ocean, for instance, which attracted tourists to the beaches, was, for the utility company, a source of coolant and what engineers called "the ultimate heat sink." To oversimplify a complex process, the generating station would use cold seawater to condense steam that ran turbines and produced electricity. Seawater would enter the plant, cool the steam, and then return to the ocean about forty degrees warmer as a result of the trip. The salt water would not become radioactive or alter chemically, nor would it diminish in volume; it would simply change temperature. Even though the plant as first proposed would require about 800,000 gallons per minute, the effect on the North Atlantic would be negligible.

Maybe, June Jones asserted, lawyers and engineers would notice no difference, but she would. She had fished out of Hampton Harbor for forty years, she told the Atomic Safety and Licensing Board (ASLB) in the summer of 1975. And the ocean, for her, was no "ultimate heat sink."

I have seen the baby herring spawn . . . by the billions out there. We . . . sat down one day in the boat [and] . . . I said, look at the water, right there in the water, a whole area was being born then, of baby herring, baby fish. . . .

But the "atomic plant" and its builders, she protested, "will kill fish."

Anything that lays an egg in the line of fish, they will kill. . . . They are going to take this good, healthy, clean ocean water that is alive with baby eggs, . . . they are going to pull it up into this monstrosity, . . . and they are going to put it back out as dead water.

Now, nothing is going to live in the water they put back out into the ocean. The water is sterilized, no life will live in it. . . .

She was so sure of the destruction she depicted that she slipped into the past tense, as if her fears had already materialized. Not only had construction killed fish, but it had also destroyed their habitat. Where, Jones asked, was the vegetation in which fish used to hide?

Now, kelp at the height and best of it is a beautiful piece of grass to see, it grows as tall as trees. Now it used to be, it would come up at low water around the rocks, above the river[;] you could see it floating over the water, . . . it was so tall. It was just like trees; a forest, that is what it was, a forest.

The ocean, for June Jones, was a beautiful forest, full of life; the "atomic plant" was "nothing more than a nightmare."[2]

Federal licensing boards prefer facts to visions, even when those visions have a factual base in experience. And the facts about a dynamic ocean environment are less than certain and open to differing interpretations. Water does not always have the same temperature or oxygen content, salinity, or acidity, and it does not hold still to be studied. Thus a description of conditions at other nuclear plants may permit comparison, but it cannot substitute for site-specific data about the local water itself. Tides, winds, seasons, currents, and a host of other factors interact to create the unique environment June Jones so obviously wanted to preserve.

The National Environmental Policy Act (NEPA), passed in 1969 and strengthened through subsequent amendment, was the legislative monument to the concern of people like June Jones. The act, and the Environmental Protection Agency (EPA) that administered it, became a favored whipping boy of the nuclear industry when the regulatory process for nuclear plants broke down during the 1970s. Frustration was certainly justified, and there was no shortage of blame to distribute, some of which environmental protection and its friends deserved. Yet the central purpose of the legislation, after all, was not the quick licensing of nuclear plants but the protection of the nation's air and water. That purpose imposed requirements, and inevitable delay, on those who wished to build nuclear generating stations.

To begin with, the cooling system Seabrook's builders wanted to use was unacceptable on its face. The EPA was supposed to keep heat, as well as more obviously toxic material, from the nation's waters. A year before June Jones decried "sterilized" water, an official at EPA had warned the regulatory staff at the Atomic Energy Commission that PSNH's design would require either an exception to regulations or another, nonpolluting method of cooling.[3]

The cooling system had caused dispute from the moment the plant was conceived. In hearings before the state siting committee in 1971, the company proposed to bring seawater two miles to the plant through pipes laid in the marsh that surrounded Hampton Harbor. Those eighteen months of hearings raised more questions than were definitively answered, but they did clarify one matter: The state would not permit pipes through the marsh, in spite of the company's promise to repair ecological damage. So, just before the record closed, professing a heightened interest in the environment, PSNH scrapped that plan and substituted tunnels beneath the marsh, the harbor, and the beach to bring coolant to the plant. (Suspicion that protection of the environment was not the sole motive spread beyond cynics and opponents of the plant; tunnels may well have been cheaper than alternate schemes, including the discarded pipeline.)

The EPA preferred the distinctive parabolic cooling towers that have become familiar at inland nuclear generating stations. Again to oversimplify, a cooling tower permits excess heat from the production of power to escape into the atmosphere; ocean water would be needed at Seabrook to promote cooling, but towers would eliminate thermal pollution of the Atlantic. A visible atmospheric plume would rise from the tower and, under certain weather conditions, additional fog and ice would plague the area; an inconsequential amount of salt would be added to the total winds already deposited on the region. The EPA believed the effect of this "closed-cycle" cooling less potentially damaging, however, than the impact of heat on the ocean from the "once-through" system.

But, the agency noted, the utility could request an exception and thereby avoid the expense and visual intrusion that towers would unde-niably create. Since design of the tunnels had begun, and since PSNH believed they would be needed anyway, the company promptly applied for an exception. It required four years, mounds of legal documents, and a couple of excursions to the courts finally to decide that the exception could be granted and that the company need not add cooling towers to already escalating costs. That result, achieved at enormous expense, was one of the few decisions that virtually every party to the dispute could accept. Those who opposed construction at Seabrook never suggested that the addition of towers would make less objectionable a facility that ought not to be built at all.

Meanwhile, on another regulatory track, the Atomic Energy Com-mission (AEC) reviewed the once-through system the utilities preferred. Hearings of the state siting committee foreshadowed a need for data about the marine environment, and PSNH engaged Normandeau Asso-ciates, Inc. (NAI), a small consulting firm, to undertake the research. Excavation at the site, NAI predicted in 1974, would inevitably cause some silting in the harbor and the marsh but would not endanger marine life. Although clam larvae sucked into the cooling system would not survive, harvests in the clam flats—regarded as the best on the state's short coastline—had been declining for years anyway. Future harvests appeared to depend not on local clams but on a floating band of larvae offshore, which wind, tide, and current would deposit in the Hampton–Seabrook estuary. Fish trapped in the tunnels would perish, but NAI's survey of the region's fish suggested the loss would be tolerable for both recreational and commercial fishermen. Officials at the AEC sent the material off to the Oak Ridge National Laboratory, where it was incorpo-rated in a Draft Environmental Statement (DES) released in April 1974.

The DES, which presumably had the AEC's imprimatur, stirred a storm of criticism even before it formally appeared. B. E. ("Bud") Barrett, a marine biologist in the New Hampshire Fish and Game Department, fired off a sarcastic five-page letter to the NRC. Research, interpretation, logic, and conclusions, Barrett charged, were all equally flawed, and terms like "hard to swallow," "grossly incomplete," "lu-dicrous," and "worthless" rattled off every page. The survey of finfish, he asserted, rested on the count from one gill net set for fourteen nights; the lobster count derived from one individual's use of fifteen or twenty traps. Even if the data were better, the fact that PSNH supervised their collec-tion was comparable to "asking the fox to guard the hen house." And the report's conclusion amounted to an admission that no one had adequate knowledge of present oceanic circumstance, coupled with a plea that the utility be allowed "to find out what went wrong after the fact."[4]

But, setting a pattern that would become distressingly familiar to the plant's opponents, the DES reflected neither Barrett's critique nor the comments of others who wondered about the quality of NAI's work. Massachusetts' commissioner of natural resources noted mildly that the data seemed a little sparse and enclosed two pages of questions to try to elicit more. The attorney general of the Commonwealth remarked that the utilities appeared to have made a premature commitment to the site and now were struggling to find enough evidence to justify a license. The Seacoast Anti-Pollution League (SAPL) endorsed the remarks of a marine biologist, who was surprised to find no mention of scallop larvae; that omission, he supposed, "may reflect more the competence of the applicants' consultant than the state of the ecosystem." Arthur Newell, another biologist in the Fish and Game Department, thought the data on clam larvae deficient also; the count had been made in a season when NAI ought to have known clams did not reproduce.[5]

In fact, the first results Normandeau Associates produced were something less than polished. The NRC staff itself, which relied on, and rarely criticized, research provided by firms applying for licenses, noted some confusion in the NAI work on currents. Further, the staff discarded as essentially irrelevant an attempt to predict the behavior of heated water discharged into the ocean; the model simply did not resemble operating circumstance. Not only were there gaps in the research design, but there was in addition an unseemly rush to print that exposed those and other flaws to scrutiny. Twice, the utilities had to send extensive corrections of the firm's work to the AEC: Some of the errors were trivial, but the necessity to send new graphs and tables and several typed pages of errata did not inspire confidence in NAI's craftsmanship.[6]

Yet, in a sense, neither the research—which improved significantly during the next few years—nor the criticism mattered. Everyone agreed that all the organisms caught in the cooling system would perish, though the NRC found less direct words to report that result. Dispute arose not from fact but from interpretation and judgment. Would mortality be so high, for instance, as to require relocation of the plant? Since the AEC had already reviewed PSNH's choice of site, clearly the agency would not require another location simply because of sloppy preliminary research; time and money would cure that deficiency. Further, evidence from a nuclear plant in California suggested that a small design change would diminish the slaughter of marine life. Reduced water velocity at the intake and a different cap on the intake shaft, the staff concluded, would bring the fish kill to an acceptably low level. These conclusions, contained in the Final Environmental Statement (FES) in December 1974, did not directly respond to any criticism and were based on no site-specific research. Similarly, the staff adopted, without much evidence, NAI's postulate that a broad band of clam larvae floated offshore; at worst, the cooling system might reduce that band by 5 percent, the staff estimated. The guess may have been reasonable, but it was not empirically based, and it did not rest on NAI's work. [7]

Thus both the criticism and the research were empty gestures; the staff could have written the FES—and concluded that the plant could be built—without either. Indeed, one of those who worked on the report testified that, "as a matter of . . . policy," there was essentially no change from one version of the document to the next. Criticism was simply reprinted in an appendix. [8] At this early stage of the proceedings, the public did not yet realize the frustration, and perhaps even futility, of participation in a system that churned out reports, based on admittedly inadequate data, without concern for criticism. In the long run, that unresponsiveness harmed the agency, the utilities, the public, and the prospects for use of nuclear power.

If the NRC staff did not seem much interested in research on the marine environment, Dr. Ernest Salo was. A professor of fisheries at the University of Washington, Salo was one of the three-member Atomic Safety and Licensing Board assigned to assess PSNH's application for a construction permit. During endless hearings, mostly in the summer and fall of 1975, the board considered the environmental reports and the testimony of witnesses interested in them. Salo was a technical expert, and he displayed both his interest and his expertise when witnesses talked about the effect of the facility on the ocean.

The witnesses were overmatched. Salo's questions exposed gaps in the staff's review of the relevant scholarship; gradually, they also revealed Salo's lack of confidence in the data the staff presented, which were essentially what NAI had collected. He challenged inferences and, in effect, the professional competence of some of the witnesses. He wanted reasons for the proposed location of the intake tunnel and discovered that no one appeared to have thought about placement that might be less environmentally sensitive. He wondered whether PSNH had plans to mitigate the alarming decline in the clam population and was told that the company thought it had done a useful service simply by documenting a phenomenon, which, of course, the proposed cooling system would accelerate. Salo's questions were polite, patient, and thorough. And he clearly was not satisfied when he finished.

Months later, when his two colleagues were ready to allow a construction permit, Salo's doubt persisted and he dissented. There was, he thought, no need for haste, because New England had generating capacity to spare. Investigation of other sites, where harm to the marine environment could be minimized, had been inadequate. But the probable harm at Seabrook, he thought, was obvious: Operation of the proposed plant "would cause a sufficient adverse impact on the aquatic biota, of commercial and recreational importance, so that other alternatives should be sought." All of the marine life of the region—clams, lobsters, fish, plants—would be harmed, and Salo did not find comfort in the small numbers advanced in the staff's reports: "I find the estimated 3% to 5% mortality rate unacceptable"; "it is my opinion that no mortality of larvae should be allowed."

That judgment led Salo to conclude that Seabrook was simply the wrong place for a nuclear power plant: Costs were too high and benefits too few. "The Applicants' approach," he wrote, relied on the assumption that "the benefits of electricity are priceless," a notion that would permit plants to be located at the whim of their owners and that "defies common sense." Nor did Salo spare the NRC staff, whose balancing of costs and benefits he did not find helpful. So, finally, he did it himself and concluded that, "I do not believe . . . it is to the best interest of the nuclear energy program (or any other energy program) to take a site that is not suitable and 'backfit' it at all costs."[9]

Salo wrote in dissent, and his two colleagues not only approved the Seabrook site but explicitly forbade the use of cooling towers, which might have reduced Salo's worry about harm to the marine environment. No explanation for the majority's view was forthcoming, a characteristic omission in a decision that provided little explanation for anything. During the hearings, when the board hinted that it might distinguish between construction with and without towers, Thomas Dignan urged that the distinction be avoided. Investment in construction on the basis of such a finding, he said, would be risky, since a central part of the design would remain unsettled. Further, he predicted, the distinction would never survive appeal.[10]

Dignan's prediction was correct, though it took about three years to demonstrate his accuracy conclusively. Meanwhile, construction, maritime research, and the regulatory process went irregularly forward. Normandeau Associates collected new and better information about fish, currents, water temperature, and the distribution of larvae and submitted reports, tables, graphs, and charts in volumes that were probably widely unread. Since the applicants paid for the research, the data predictably showed a tolerable environmental cost. The Environmental Protection Agency gave leisurely consideration to the cooling system, required relocation of the intake and the diffuser, changed its mind once or twice, was sued and required to reconsider, and finally provided a judicially acceptable certification almost four years after the official who presumably had the authority to decide had given his first tentative approval. And the tunnels, eighteen feet in diameter, each more than three miles long, crept through the rock more than 200 feet below the surface to a destination more than a mile and a quarter offshore.

Those tunnels were impressive aqueducts. Each reactor would require 412,000 gallons of coolant per minute, which would flow at a rate of one foot per second into one of three intake shafts about sixty feet beneath the water's surface. A "monitor cap," more than thirty feet across, covered the ten-foot shaft; similar structures elsewhere seemed to discourage curious fish, which were not prevalent at that depth off Seabrook in any case. The water would descend about 175 feet to the tunnel, flow downhill to a point almost 250 feet below the plant, and arrive about forty-four minutes later at the pumps. After cooling the steam, the seawater, now about forty degrees warmer, would require another three-quarters of an hour to return to the ocean through the other tunnel, which ended in twenty-two nozzles, about ten feet off the ocean floor and fifty to sixty feet underwater. After the heated water diffused, models showed that the surface temperature might rise three or four degrees across an area of perhaps fifteen acres.

If that result seemed environmentally unimportant, and therefore fortunate, the system had changed substantially in the years following Ernest Salo's admonitions. And those changes, which were surely improvements, derived from the tedious licensing process about which everyone complained. The continuing research of NAI, which went forward while the litigation went on, increased the quality as well as the quantity of available information. As more became known about currents, the behavior of particular species of fish, the development of clam larvae, and the nature of the water column from surface to floor, the cooling system was redesigned. The intake was moved away from the clam larvae that might be carried into the estuary and placed at a depth where fewest fish would be trapped; designers tinkered with monitor caps and diffusers to minimize the effects of heat and current on various forms of marine life. The costs of delay and modification were high—certainly too high in the view of the utilities. But none of those improved components would have been built in the same way if the licensing process had been quick, efficient, and uncontested. "Let's not kid ourselves," Thomas Dignan had remarked in 1974. "We are going to go through this cooling system brick by brick, pipe by pipe, fish by fish, and we might as well get at it." It was not bad prophecy. Nor was it a bad idea.[11]

The tunnels at the heart of the cooling system were an afterthought, which Dignan had hoped in 1972 might satisfy the requirements of the state's laws protecting wetland and pacify those who were opposing the plant in the state siting hearings. That hope, which appears naive in the light of years of subsequent legal wrangling, suggests Dignan's estimate of the importance of the marsh the tunnels spared. The original idea that PSNH would cut a ditch, fifty-five feet in width, through the estuary that separated the plant from the ocean was clearly intolerable. But the new plan, which recognized the importance of the marsh and required no major construction there, might, Dignan thought, end controversy.

Defense of the marsh, chronologically and symbolically the first cause of opponents of Seabrook Station, rested not only on the state's statutes but also on an informed understanding of the environmental functions of wetlands. A broad popular effort, based on this environmental consciousness, had prevented location of an oil refinery, associated with shipping tycoon Aristotle Onassis, on an estuary north of Seabrook. Grass-roots groups, connected to the political process through local leaders, challenged Governor Meldrim Thomson, a vigorous proponent of the proposal, and secured a state law allowing local communities to veto such facilities. Given the opportunity, Durham voted to bar a refinery, and Onassis's tankers sailed away. But New Hampshire's cherished tradition of home rule did not apply to federally regulated nuclear power plants; subsequent votes in Seabrook declining to accept Seabrook Station, though interesting, were quixotic.

Yet to mobilized environmentalists, fresh from a triumph over an international magnate of fabled wealth and influence, the oil industry, and the state's political establishment, stopping Seabrook Station seemed both an obvious and a possible next step. That estimate, of course, was strikingly mistaken. In retrospect, neither Onassis nor the oil cartel had much stake in a New Hampshire refinery, nor had they any desire for controversy. The nuclear industry, in contrast, was seriously worried about a diminished stream of orders for plants that were beginning to encounter effective resistance. And the federal regulations of the Nuclear Regulatory Commission were much less responsive to local protest than were the laws and legislature of New Hampshire. Finally, PSNH, a relatively small utility when it first sponsored Seabrook Station, came to see the facility as indispensable to its corporate future. That attitude energized lawyers and lobbyists and an enlarged staff in public relations.

Opponents challenged PSNH's assertion that construction would not harm the marsh. A new road to the site from U.S. Route 1, a barge landing built into Hampton Harbor, excavation for containment structures, construction of switching yards and transmission lines, to say nothing of miles of tunnels, seemed to environmentalists likely to disturb an ecosystem they usually described as "fragile." The Audubon Society, which owned part of the marsh and wanted to retain title, argued that wetlands were vital resources for waterfowl. Construction would disrupt migratory habits, the society held, and power lines and huge structures would maim or kill large numbers of birds. The Seacoast Anti-Pollution League predicted that excavation would result in silting in the estuary and harm marine life. The Society for the Protection of New Hampshire Forests (SPNHF), one of the largest landowners in the state, had little direct interest in the salt marsh but a major concern about plans to string transmission lines across a freshwater marsh the society owned nearby. The army's Corps of Engineers, not noted as a friend of nature, warned PSNH not to let mud from excavation or the barge landing stray into the harbor, over which the Corps claimed jurisdiction.

The utility tended to respond with calm reassurance to all these concerns, a stock reply that persuaded only the NRC. Staff members testified that no one could predict with certainty how much silting would result from dredging a channel to the barge landing, but the effect was "not expected to be severe."[12] The basis for that comforting expectation was not apparent. Indeed, when the Corps of Engineers reviewed environmental documents prepared by the company and the NRC, they seemed so deficient "that we must explore ways of supplementing" them or else "start from scratch" to prepare an environmental statement the Corps could support. The Corps invited NRC representation at a public hearing to be held in Seabrook just before the construction permit was issued in the summer of 1976. Pleading a fine legal point, the NRC declined to appear.[13]

The major concern of the Corps of Engineers, as construction progressed, was the muddy drainage that found its way to the waters around
the site. Construction required excavation, and dirt became mud when
the hole reached the level of groundwater or thunderstorms rolled over
the region. Turbidity—that is, soil suspended in water—was a threat to
marine life, especially lobsters, and the company and its contractors
were supposed to control the problem. The first effort toward that end
was no fancy engineering device but a traditional New England remedy:
a dike of baled hay. It was a disarmingly simple solution, and it did not
work. Nor did a settling pond, where pumps were supposed to route
water from the excavation, completely end the muddying of nearby
waters. Inattentive workers sometimes allowed mud to find its own way
off the site, a practice that provoked more than one reprimand from
inspectors. During the frenzied first burst of activity, the company
monitored turbidity from the air, an obviously imprecise system of
measurement. On the basis of one such flight, a utility official testified
that there had been some "minor" runoff, a remark that left him
vulnerable to pointed questions about how "minor" runoff was that
could be observed from a moving helicopter. [14]

When construction was nearly complete, PSNH summed up the
environmental disturbance, which, like everything else about the project, exceeded the company's expectations. Heavy traffic and consequent
congestion on U.S. Route 1, a major north–south highway, required
addition of a second access road to the plant. Since the town of Seabrook
could not supply sufficient fresh water, the company had had to drill two
wells. Although first plans called for preservation of trees to serve as a
visual screen at the boundary, some of them had to be cut. More land
than anticipated had been required to store equipment and rock removed from the tunnels. The company attributed errors in prediction
"to the construction schedule which was delayed and perturbed by the
state and federal licensing and rate-making process." The explanation
was gratuitous, and the mistakes, given the scope of the project, essentially harmless. If other estimates of PSNH had been no worse, the
company's troubles would have been insignificant. [15]

The temptation to use bureaucratic delay to respond to all criticism and
answer most questions would prove overwhelming as deadlines went by,
costs climbed, lawyers sparred, and no current flowed. The nuclear
industry, unable silently to endure cancelation of contracts and scrap-

ping of proposed plants, sometimes suggested that the NRC rewrite its rules to reduce or even eliminate public participation, a step that would undeniably accelerate the issuance of licenses. The public ought to rely on the NRC, one executive asserted, to protect the health and safety of the population. "But to have housewives coming into these highly sophisticated technical discussions," he continued, "is just ridiculous."[16]

She was a housewife, but nobody called Elizabeth ("Dolly") Weinhold's intervention "ridiculous." She had participated in the state siting hearings, and, late in the summer of 1973, she asked the AEC for formal standing as an intervenor at the federal level. Her petition was not couched in the usual legalese, bore no citations, contained no references. She wrote a simple, first-person testimonial about children, clams, and surf. But she realized that intervention could not rest on the recreational importance of the coast to her family, so she learned a great deal about the geology of the region; she wanted the NRC to share her apprehension about earthquakes. Weinhold knew that access roads to the beach had been built on filled marsh and that any earthquake might make them impassable when they were most needed. If that happened, people would have "to swim or fly" to leave Hampton Beach.[17]

Thomas Dignan wrote a detailed brief opposing Mrs. Weinhold's intervention. She read the document carefully, used it to correct the legal defects he had identified, and thanked him sincerely. Indeed, Dignan and other lawyers in the case often assisted her with legal niceties and were ordinarily patient with and tolerant of her legal lapses. She showed more interest in substance than in the form of her questions during cross-examination of Dignan's witnesses, for instance, and he did not press objections that might well have been sustained. When she wanted to subpoena a witness from Colorado at the NRC's expense, the board granted the request, though that ruling seemed to contradict both policy and precedent. When she lacked information about costs, the assistant attorney general of New Hampshire volunteered to find out for her. She acknowledged graciously the help she received, and she pressed on: "I want to thank all you gentlemen for the patience you have had with me on my crash course in legal techniques," she said one morning; "I do hope I do better today."[18]

The assistance was offered without condescension because Dolly Weinhold was in fact a full colleague in the licensing hearings. She presented witnesses and asked questions, and she had done her homework in preparation. She wrote detailed letters to members of the regulatory staff, disclosing the sources and results of her research on the geologic structure underlying the site. She read accounts of eighteenth-century earthquakes, two of which were the largest disturbances on record; she related those events to the geology of the region and to subsequent tremors, which she carefully tabulated. At a meeting of the Advisory Committee on Reactor Safeguards (ACRS), she noted that the company's seismic consultant had disparaged a source at one point in his argument while accepting without question data from the same source elsewhere: "in other words," she said, "they are using" the source "when it benefits them." She had to frame questions during the discovery phase of the hearings, and she had to write responses to the queries of lawyers from PSNH and the NRC staff. She had no secretary and she had no money and she had no scientific or legal training. But she could type and she could listen and read, and she used common sense and learned quickly. [19]

In particular, she studied the tunnels. Weinhold was not persuaded that the plant could withstand an earthquake of the scale she believed possible; she was certain the tunnels could not. Coolant, which would be crucial in the event of a mishap, might not be available because cement, rock, and water would react independently to the movement of an earthquake, and the tunnels might collapse or otherwise become inoperative. [20]

The NRC staff itself shared Mrs. Weinhold's misgivings. Before her inquiry, the staff had instructed PSNH to add a system to compensate for the loss of the tunnels during an earthquake. The company disagreed with the staff's assessment but added a small cooling tower in order to avoid "the extra time and delay inherent in further geological investigation and design effort," which, the company recognized, might "not conclusively prove the adequacy of the tunnels" in any case. When another intervenor wondered whether the emergency system might require so much water as to harm the marsh, Thomas Dignan momentarily lost his patience: If there were an earthquake, Dignan said, which destroyed the tunnels, damaged the plant, and moved the New Hampshire coastline "fairly near Vermont," the water level in the marsh might not be the first worry. [21]

Mrs. Weinhold was still not satisfied. Why, she asked, if the tower was intended to serve in a seismic catastrophe, was it not designed to more stringent specifications than the rest of the plant? "In my opinion," she wrote, "an emergency backup system . . . is supposed to be a fail-safe system and should be designed more conservatively than any other part of the facility." She reviewed again her case for protection from a more destructive quake than the NRC had decided was "credible." She also submitted new evidence, including supportive scholarship by David Okrent, a member of the ACRS whom she wanted to testify at the licensing hearings. Because of his relationship with the NRC, Okrent was prohibited from appearing, a ruling that simply baffled Elizabeth Weinhold. "I am still trying to figure out," she wrote several years later, while the seismic issue remained unsettled and construction went furiously ahead, "why an Advisory Committee was instituted to advise the NRC . . . when the Commission doesn't take the advice." Other experts, she had discovered, were reluctant to testify on behalf of intervenors because their research frequently depended on federal grants, usually, in the case of nuclear experts, from the NRC; "they are fearful of not having positions in the future," she said, "either with the government or with the utilities."[22]

In fact, experts did occasionally side with intervenors, including Elizabeth Weinhold. The problem was less a lack of experts than a lack of certainty in the science of earthquake prediction. As late as 1983, when the ACRS considered recent tremors in the region, heard again from Mrs. Weinhold, and solicited outside opinion, the issue remained murky. One consultant believed the NRC staff had underestimated the magnitude of earlier quakes in New England and therefore imposed too few safety requirements on the utility; another confirmed the staff's work in every respect. William Kerr, chairman of the Seabrook subcommittee of the ACRS, was able to joke about the obfuscation that cloaked uncertainty. He asked what might happen to one of the plant's systems in the event of an earthquake and received a long, detailed, qualified, and ultimately blustering response. A pensive Kerr then asked the witness if he had ever heard "the story of the little boy who asked his father where he came from." The witness had "not really" heard it. "Well," Kerr said, "his father decided it was time to have that heart-to-heart talk and when he finished his long story the little boy said, 'Well, my friend up the street came from Boston, and I just wondered where I came from.'" Now, Kerr continued, thinking of the wordy, defensive answer he had just heard, "What I want to know is about Boston." The "Boston" answer was that the witness did not know.[23]

The seismic controversy lingered in the Seabrook litigation largely because of the views of Dr. Michael Chinnery, head of the Applied Seismology Group at the Massachusetts Institute of Technology's Lincoln Laboratories. He relied on mathematical probability, in addition to geology, to predict seismic disturbance. Chinnery's method was controversial but intriguing, and the NRC, at various levels, considered his work on so many occasions that Dignan finally compiled a chronology in 1981 showing the number of times Chinnery's hypotheses had been dismissed; Dignan made his list too early, for there were additional reconsiderations with the same result. The majority of the appeal board gave Chinnery's views a searching critique and rejected them in 1977, but a dissent found them persuasive. Faced with this sort of uncertainty, the NRC hesitated and ordered the matter reexamined once more; not even Mrs. Weinhold could complain about the time and attention the topic received.

Chinnery himself was no foe of nuclear power. He did, however, urge the NRC to weigh his conclusions and methodology against more conventional seismic science. When that investigation concluded, he predicted, "the Board will find . . . that we know remarkably little about earthquakes in the New England area." The problem, in Chinnery's view, was "not to say who is right and who is wrong, but rather to define a conservative estimate for the Safe Shutdown Earthquake given this lack of knowledge." The NRC never held the scholarly shoot-out Chinnery recommended and eventually simply decided to pick a point of view and endorse it. Predictably, the view was that of the utility and the NRC staff.[24] That decision did not resolve the underground ambiguities, but it did allow the plant to rise above them.

Elizabeth Weinhold's apprehension about earthquakes would have diminished considerably had PSNH decided to use another fuel to generate electricity at Seabrook Station. Whatever the pretext, the prospect of nuclear fission, and the real or imagined perils of that technology, stimulated much of the fear and passion that sustained resistance to the project. A spirited defense of birds or fish or mollusks, for instance, also served as a way to protect people from the hazards of radiation. If the law seemed to erect more safeguards for wildlife than for humans, the legislatures, environmentalists argued, erred in emphasis. Dedication to environmental preservation was no less sincere because people, who shared the environment with other forms of life, would be afforded the same protection other species enjoyed. Preservation of the marsh, a worthy objective in itself, might also serve to block the spread of nuclear power. Many of the causes championed by Seabrook's opponents, therefore, had both an immediate goal and the indirect objective of avoiding all of the dangers associated with nuclear fuel.

The Society for the Protection of New Hampshire Forests differed in this respect from other environmental groups opposing Seabrook Station. Established about a half-century before the atomic age began, SPNHF had actively promoted the establishment of wilderness areas in the state. The organization did not depend on car washes and bake sales for its considerable budget but owned a great deal of real estate and had close ties to the business and political leadership of the state.

A decision to intervene was not easily taken, for many members of the organization, out of habit and association, accepted without hesitation the utility's word about the safety and necessity of the proposed plant. The society took no stand on nuclear power, which would have divided environmental militants from members who had traditionally given the organization influence and support. And, in effect, the focus of SPNHF's intervention on the single issue of the route of transmission lines conceded that the plant could be built at Seabrook. Unlike several other intervenors, then, SPNHF did not attempt to reopen every previously settled question at every subsequent opportunity.

Thomas Dignan argued that questions about the transmission lines had been settled too. The location of high-voltage lines, he claimed, was a matter for state jurisdiction and had been resolved by local authorities, whose familiarity with the region allowed a judicious balancing of competing interests. Federal officials, on the other hand, would have either to spend inordinate time learning local circumstance or else to ignore it and make mistakes. The argument served his case and was not out of place. But Dignan must have felt vaguely uncomfortable making it, because he usually sought to preempt local control in the interest of his client and the federal NRC.

To move the power from Seabrook Station to the New England electrical grid would require about fifty miles of new high-voltage lines in three New Hampshire corridors. Two of those corridors, as proposed, crossed inland wet areas in which environmentalists had interests; one of the two cut through land owned by the Society for the Protection of New Hampshire Forests. The state declined to interfere with PSNH's plan, inaction that explained Dignan's desire to keep the matter out of federal proceedings. But the NRC had to assess the environmental impact of the whole project, which included, of course, provision for electricity to leave the site.

The utility's preferred route from Seabrook south to Massachusetts crossed the Great Cedar Swamp in Kingston, which SPNHF owned. Avoiding the area, the utility claimed, would not only add $500,000 to the cost of construction but would also diminish the reliability of the system. The half-million dollars seemed larger in preconstruction discussion than would be the case when escalating overruns made that sum look like small change. Whatever the cost, SPNHF argued, it was worth paying to save an unspoiled, freshwater area that had unique environmental and recreational importance. The society mobilized local environmental groups and its own members in a letter-writing effort that made the southern transmission corridor seem the most important environmental question Seabrook Station presented. A limited issue of that sort could be compromised, and the NRC could afford to pay attention to its mail.

PSNH too offered to compromise. The utility would use enormous poles to suspend wires over the Atlantic cedars that were the major concern of the society. This proposal would have reduced nearly to zero the number of trees to be cut, fewer than would be taken if the NRC's suggested route skirting the edge of the swamp were followed. SPNHF pushed for a longer detour yet, which might have required condemning a residence or two but would have spared trees and kept the wires out of sight. In the Final Environmental Statement (1974), the staff balanced the visual intrusion against "more environmentally sensitive" concerns and specified an alternate corridor, called in subsequent litigation "the staff's modified dogleg," around the swamp. The compromise satisfied neither the utility nor SPNHF. But an opportunity to demonstrate environmental sensitivity and to react affirmatively to constructive popular criticism, without incurring significant cost and without jeopardizing an important principle, came infrequently. The staff responded.[25]

That victory effectively took SPNHF out of the case, but the march of towers across the landscape kept the issue alive for others. Several towns required PSNH to bury wires in ordinances New Hampshire courts did not enforce. When PSNH applied for an operating license in 1982, residents of South Hampton tried to persuade the licensing board to reconsider the location of the transmission corridor to Massachusetts. Organized as the Society for the Protection of the Environment of Southeastern New Hampshire and supported by town officials, these critics advanced several bases for their protest. The towers, selectmen of South Hampton noted, exceeded the height the town's zoning ordinance allowed for any structure in the community. In addition, the corridor adversely affected an officially designated historic district, which the law on historic preservation appeared to forbid. Finally, people living in the vicinity of high-voltage lines wondered whether that proximity would endanger their health, and they were not ready passively to accept the staff's dismissal of their concern.

The staff, wrote Charles Goldstein, relied on research funded by agencies, including the U.S. Navy and several utilities, with a direct interest in a benign finding. Other scientists, including some in the Soviet Union, had positively identified physiological results of prolonged exposure to overhead high-voltage lines. Investigators for the Public Utilities Commission of New York State and the American Cancer Society, Goldstein said, had confirmed that work. The NRC staff appeared to be outside the scientific consensus. But PSNH and the NRC staff evaded Goldstein's challenge and others to the transmission corridors after 1976. When a new query arrived, Dignan fired off another brief maintaining that the NRC's construction permit had definitively settled the matter. He was invariably sustained. [26]

Consideration of another transmission corridor exposed incompetence as well as the partisanship of the NRC staff. The National Environmental Policy Act required preparation of a cost/benefit analysis comparing Seabrook to other possible sites. Neither PSNH nor the NRC staff took the provision seriously, and the appeal board eventually characterized the first pass at the task as "patently insufficient," which was not the sort of language NRC boards usually applied to the agency's staff.

The criticism might have been even sharper. Dr. Robert Geckler of the NRC staff, assigned to compare costs of distribution of power from Seabrook with those that would be incurred at a site on the Connecticut shore, calculated that the utilities would have to spend an additional $54 million to transmit electricity to PSNH's New Hampshire customers. That cost, Geckler said, would be avoided at Seabrook, because existing lines would carry Seabrook power to the Connecticut customers of PSNH's minority partners. The incremental cost Geckler charged against Connecticut sites puzzled Donald Stever, an assistant attorney general of New Hampshire. Why, he wondered, would electricity produced in Connecticut for sale in New Hampshire require transmission lines that were not needed if power generated in New Hampshire were sold in Connecticut? Geckler's fumbling response to Stever's query revealed that the NRC's environmental expert did not realize that electrical energy could flow through existing lines in either direction. As he grasped the dimension of Geckler's ignorance, an incredulous Stever asked the question directly: "if the power can flow south in the grid, do we assume it could also flow north?" Geckler did not know. John Frysiak, chairman of the ASLB and no friend of intervenors, asked the flustered witness whether transmission lines did not already link Seabrook with distribution facilities in Connecticut. Geckler said he did "not know the grid" and did not understand how either nuclear plant would "tie into it." Geckler's testimony and the documents he had prepared obviously provided an insecure base for the requisite cost/benefit analysis; Stever moved to strike the testimony.

But the staff's embarrassment was not complete. Marcia Mulkey, counsel for the staff, opposed Stever's motion, defended Geckler's work, and thereby indicated that she comprehended neither the operation of an electrical grid nor the devastating cross-examination of her witness. She wanted Geckler's testimony in the record, in spite of Stever's exasperated comment that "it is all fiction."

> If the board is in the business of making decisions on the basis of hypothetical fiction [Stever continued], produced by witnesses who do not understand what is going on, . . . then fine, it can stay in. But . . . if the testimony and the record is to be at all meaningful, it has to come out.

Frysiak broke the long silence that followed with a ruling that Stever's objection was well made. It was, Stever wrote afterward, "one of the worst performances by a government agency to take place in a public forum."[27]

Stever's indictment might have been extended to the rest of the process of evaluating alternate sites. The comparison was never a task for which PSNH had any heart, since the utility's choice of Seabrook was a considered managerial judgment, not capriciously made. The NRC paid no attention to public reservations about the site and little more to qualms expressed by its own staff. In 1973 Harold Denton, destined to become a major figure in the regulatory hierarchy, wrote Joseph Hendrie, who would later become a commissioner, that the plan for Seabrook threatened the marsh, "a unique natural resource that should be preserved." In addition, the transient population on the beaches and the rapid growth of the region might make another location preferable; Denton suggested a couple that seemed promising. Even though PSNH might cure some of the deficiencies with money, Denton's review indicated "that there are several realistic alternatives to the Seabrook Nuclear Station if costs of delay are not allowed to be the controlling factor."[28]

The NRC was under constant attack for holding back the wave of nuclear progress, and Denton's reference to the "costs of delay" gave Seabrook an immediate edge over any other site. A few months later Denton himself received an internal memorandum from the environmental specialists in the agency. They concluded, as he had, that the case for Seabrook was not entirely persuasive, and they were disappointed when their reservations about the site did not appear in the environmental statements.

> The statement that Seabrook is the best choice can hardly be made from the viewpoint of an indepth analysis. Such an analysis might yield "with proper safeguards, acceptable." The same, if not more, can be said for other locations. [29]

Right from the start, the unsettled question of the cooling system, and the jurisdictional confusion surrounding it, plagued the analysis of alternate sites, which in turn led to two suspensions of construction, two years of bureaucratic scuffling, and a shelf full of legal documents—all of which made no substantive difference. Before the charade came to a merciful end, lawyers had taken to citing *Through the Looking-Glass* as well as NRC regulations, since some of the precedents from the Mad Hatter's tea party seemed entirely too relevant.

As argued by Thomas Dignan, the position of PSNH was simple and consistent: The company had made an informed decision after studying other locations; none was superior to Seabrook. Investigation of any other site would be costly, time-consuming, and not the company's responsibility, since PSNH could hardly be asked to develop plans for nuclear plants all over New England in order to compare potential costs and benefits to those of Seabrook Station. Further, the utility lacked the power of eminent domain outside New Hampshire, so acquisition of land, even of preferable land, could not be assumed. And finally, although expenditure of funds for construction was explicitly at the company's risk, that expenditure ought to count in a comparative analysis; dollars spent, at no matter whose risk, were costs that ought to be acknowledged. [30]

The NRC's regulations did permit consideration of those "sunk costs" and required that an alternate site be "obviously superior" to the one the utility preferred. That standard became increasingly difficult to meet as a great deal of PSNH's money rapidly became "sunk costs" in the excavations and tunnels of Seabrook. Yet in January 1977, uneasy about the staff's slipshod analysis of alternate sites even before Geckler's disastrous testimony and concerned about a still unapproved cooling system, the appeal board ordered construction suspended. Those doubts outweighed the utility's sunk costs for a majority of the panel, which concluded that "it makes no sense now to proceed," because there was a "manifestly real" chance "that the site will ultimately be rejected in favor of some alternative."[31]

On appeal, the NRC decided that construction and site comparison could proceed simultaneously. The Commonwealth of Massachusetts understood the commission's desire to avoid the costs, hardships, and inefficiencies that would accompany a temporary suspension. But that practical motivation, Assistant Attorney General Ellyn Weiss noted, prejudged the case, for it assumed suspension would indeed be temporary and that work would eventually resume at Seabrook. All that activity, after all, had no point unless a nuclear reactor was ultimately going to generate electricity; thus the NRC's presumption that suspension would only constitute an interruption robbed of any significance the still incomplete review of alternate sites.[32]

Significant or not, review ground inexorably onward. The NRC assigned tasks to both licensing boards and predicated final approval of Seabrook Station on an affirmative finding, after comparison with other sites, that the facility was acceptable with cooling towers, unless the EPA ruled in the meantime that towers were unnecessary. In making the comparison, the NRC noted, the ASLB could take into account expenditures already made and the additional cost of delay, an instruction that provided the licensing board with unmistakable direction. Then, perhaps sensitized by considerable criticism from, among others, New Hampshire's political leaders, the NRC gave a second clear hint. The failure of the several regulatory boards to complete prompt, decisive, and, especially, final action had made Seabrook a notorious "paradigm of fragmented and uncoordinated government decision making on energy matters." The commission had good ears, and there could have been little question about the judgment subordinate boards were supposed to render. But, pending the review it ordered, the NRC affirmed the appeal board's suspension of the construction permit, which hardly calmed the critical uproar.[33]

The burden of the commission's order was that licensing boards ought to streamline procedure, complete their approval of the project, and turn on the current. But the licensing "system" was not hierarchical, intervenors did not genuflect when the NRC decreed, and no one had much interest in the evaluation of alternate sites that the NEPA required. Intervenors, some of whom opposed nuclear power on principle, disclaimed any obligation to locate a plant they hoped would never operate. PSNH argued, as already noted, that the applicants had completed their survey of alternate sites before proposing Seabrook. That appeared to leave the task to the NRC staff, which had neither time nor inclination to make a detailed, thorough investigation of every possible place in New England that might prove superior to Seabrook. And, to judge by its earlier efforts, the staff lacked expertise as well. The commissioners' order seemed curiously detached from the bureaucratic process for which they were presumably responsible.

Lacking any choice in the matter, the staff dutifully turned to the task. The commission's sense of urgency had obviously registered, for a new batch of "expert" testimony appeared within a few weeks. This analysis, the staff avowed, was "independent" of PSNH, an assurance that seemed to make a virtue of the utility's grudging and incomplete cooperation.[34] The staff limited its assignment by rejecting out of hand sites outside the service area of the utilities involved. Then investigators "made a reconnaissance level review of the environmental and economic costs of replacing Seabrook with additional units at existing nuclear stations" and once more checked suggested sites in northern New England. Cross-examination revealed that "a reconnaissance level review" consisted of a brief visit or, in some places, a helicopter flight over the proposed site, an investigatory technique Ernest Salo deplored. He was particularly interested in a New Hampshire site that seemed to him more promising than Seabrook. But the staff's visit had lasted "about 15 minutes" and provided no basis for him to make a judgment.

When the staff concluded that Seabrook was acceptable even with cooling towers, Salo dissented, as he had from the initial decision. "If cooling towers are acceptable at Seabrook," he wrote, "they are acceptable" anywhere, and the requirement that a plant must be aesthetically acceptable should be removed from the NRC's rules. Addition of towers would constitute "the ultimate . . . backfitting of a site." But, late in 1977, Salo's colleagues accepted the staff's findings on alternate sites as well as towers, and the show moved once again to the appeal board.[35]

Since the staff had provided nothing new on other locations and examination quickly revealed how shallow the investigation had been, lawyers concentrated on the conclusion that Seabrook could be constructed with towers if necessary. That assessment placed PSNH in the awkward position of defending a plan the company had not advanced and did not like; indeed, intervenors put on the record a broadside PSNH had distributed eighteen months earlier describing a parade of horrible results of cooling towers. A tight-lipped PSNH witness said the company no longer adhered to that view. When Dignan tried to introduce photographs of the site to which an artist had added towers in an attempt to depict visual impact, he encountered a flurry of technical objections. So he withdrew the offer. He had, he said, "no desire to put the photographs in the record. I at this time agree the towers look like hell. We have been saying it for years. There is no doubt about it."[36] Within a few weeks, the EPA approved once-through cooling, the hypothetical towers temporarily disappeared from the dispute, and the question about alternate sites was presumably settled.

But it took another year. The courts threw out EPA's approval of once-through cooling, and repairing that technical deficiency required some months. Moreover, the appeal board found the licensing board's opinion so deficient, and the staff's work so appalling, that the precedent could not stand. The irritation and impatience of the members of the appeal board showed in the questions they sprayed at counsel for all parties in March 1978. Chairman Alan Rosenthal chided the staff for submitting an inadequately developed record and for relying on experts without expertise, an apparent reference to Geckler's testimony on the transmission lines between Seabrook and Connecticut. Michael Farrar, who had dissented some months before from the decision to reinstate the construction permit, revealed both the board's sensitivity to criticism and his own annoyance with the staff, whose shoddy work had left him without defense. If the board simply affirmed the ASLB decision,

> we are going to look like the bad guys. We will be the only stupid people in the world who don't know that these sites are not obviously superior or the transmission distance is too far. No one will understand that the reason we are so stupid is that nobody gave us information. They won't believe that happened in the most controversial case ever before the Commission.[37]

"That is not a question," Farrar concluded unnecessarily; "it's a speech."

It also foreshadowed the board's decision, which reopened the question of alternate sites and criticized the licensing board as well as the staff. The ASLB's findings on alternate sites, which paralleled recommendations of the staff and the applicants, were inadequate; the staff's "perfunctory" investigation led to an appraisal that was "unjustifiably slanted in favor of Seabrook." The cost/benefit comparison rested in part on the expense of transmission, which had been happily removed from the record. Farrar thought the "hurried, careless analyses furnished by the staff" underlay a "superficial and incomplete" decision to allow construction; he would have suspended the permit once more, but his colleagues disagreed. One of them remarked resignedly that requiring yet another alternate site analysis would have no substantive result. At considerable expense, the staff would prepare another record demonstrating that Seabrook was the best possible site. Those "further proceedings," predicted Dr. John Buck, "may thus be viewed as just another pull on the Seabrook yo-yo string." His footnote made the point explicitly: "Pulling a yo-yo string expends energy to produce rapid spinning but yields no productive results."[38]

The staff, predictably, disagreed with the appeal board's order and asked the NRC whether a full-scale investigation really was necessary, a query that brought a blazing rejoinder from Robert Backus. The staff's performance in the matter, Backus said, was uniformly perceived as "dismal." "Now, having been caught in an abysmal failure to discharge their responsibilities," he continued, "the staff comes before the Commission asking to be relieved of its embarrassment" through a change of assignment.

Backus's blast provoked a retaliatory allegation of bad faith from Thomas Dignan. The whole controversy over alternate sites, Dignan charged, had nothing to do with the environment and much to do with delay. Backus's client, Dignan claimed, "believes in more review for the sake of review," which was an abuse of legal process. Other intervenors were on record in other proceedings in opposition to the very sites they now proposed be considered as alternatives to Seabrook. It was "an inescapable fact," Dignan noted in italics, "that *no intervenor wants Seabrook moved*" to several of the sites under consideration.

For its part, the staff conceded some confusion in the face of conflicting instructions from various levels of the agency, a situation that recalled the croquet game Alice encountered in Wonderland, where everyone shouted and made up rules and "you've no idea how confusing it is." But, like Dr. Buck, the staff predicted that further review would only confirm Seabrook's advantages; on the whole, the staff said, it would prefer not to make the effort.[39] The commission decided that the staff would have to do so anyway and, in June 1978, suspended the construction permit again.

Commissioner Richard Kennedy seized the occasion to deliver—in dissent—another stinging lecture to the NRC's staff. Suspension, Kennedy said, penalized the utility for the inadequacies and errors of the staff, which had several times failed to provide an adequate analysis of alternate sites. But the costs of those errors fell not on bureaucrats but on the applicants and ultimately on those who purchased electricity. "We simply cannot risk another 'Seabrook' in the future," Kennedy wrote, and he urged, but did not specify, "major steps . . . to reform the process under which we work." Without reform, he said, "there can only be an unnecessary, but possibly endless, repetition of the debacle we have helped perpetuate and now must face here."[40] A few weeks later, when an EPA ruling provided a legal pretext for reinstating the permit, Kennedy was unpacified. Although it had been brief, he deplored the interruption in construction that "deprived people of employment." The "bureaucratic bungling" and "past procedural errors" that had caused the unnecessary hiatus "in no small measure must be laid at the doorstep of this commission and its own staff."[41]

Yet the record was still incomplete. The comparative balancing of costs and benefits that all parties agreed NEPA required had not been placed on the record, two years and two suspensions after construction had begun. It was a task for which nobody had much motivation, and Robert Backus, for one, signed off:

> We can well understand, indeed we share, the Appeal Board's frustration at being asked to perform this absurd task of alternate site review at this point. Every party to the proceedings recognizes that it is a joke. Indeed counsel for the staff had suggested to [me], *sotto voce*, that they are only going through this exercise in an attempt to sharpen up their skills for another application where the alternate site inquiry might involve something more than shadow boxing.

The commission's directive on sunk costs, Backus charged, made the "proceedings . . . a sham." Marcia Mulkey, counsel for the staff, admitted that she thought the analysis would be good practice but denied that the investigation was anything less "than a serious good-faith effort."[42]

And, in fact, at long last, that is what it was: a document, about two inches thick, stuffed with tables and data, demonstrating that Seabrook was superior to nearly two dozen other locations. One coastal site in Maine might have been marginally better than Seabrook, but consideration of the cost of delay and the sunk costs at Seabrook eliminated that edge. And the staff clearly did not want to undertake a study of similar scope ever again: "The number and volume of analyses could proliferate beyond any reasonable need if this approach were generally adopted." The staff promised to study the matter and adopt new "review procedures for other proceedings." The resolution was a trifle tardy.[43]

The appeal board played out the last act in January 1979 during a two-day hearing on the document, which, in spite of the staff's good-faith effort, was not entirely reassuring. The board quickly found errors in population, mileage, and weather data, all of which ought to have been precise but which rested on assumptions made in Washington instead of on facts collected in the field. In estimating the pool of labor that might be available at any construction site, for instance, the staff had calculated distance without consulting local police or driving the roads. To the consternation of Chairman Rosenthal, the resulting tables were in "crow-miles," or "non-road miles"; construction workers, the chairman observed, must have wings. A disgusted Dignan remarked that most of the testimony had nothing to do with the real world anyhow. If the population of the alternate site did not differ from that around Seabrook by a factor larger than two, the staff concluded that the difference was not significant. Where, Rosenthal asked, did that rule come from? "Well," the staff witness replied, "it was simply . . . a guideline we developed." What appeared at first glance to be a method of inquiry, in other words, was simply an ad hoc means to the end of demonstrating Seabrook's superiority.

Numerical data, which appeared precise, were not, and nonquantitative "facts" proved equally slippery. To probe the effect of major construction on a community's labor supply, Backus asked about unemployment in several of the places under consideration. The staff thought any differences unimportant because "unemployment rates have a history of shifting very dramatically during short periods of time." Had support for or opposition to nuclear power been gauged? No, the staff had no scheme for measurement. How about votes, Backus asked. It was, the witness replied, "very difficult to factor those votes into a cost-benefit analysis." Michael Farrar summed up: "one, it's hard to determine what the community attitudes are, but, two, even if you know . . . there is nothing in your opinion that you can do with that information." "Yes," the witness responded.[44]

The appeal board let the whole matter slide for a few months, when a court decision upholding EPA's approval of once-through cooling appeared to eliminate the need to make a final determination. It was not a very dignified burial.

The issue deserved better. A comparative appraisal of costs and benefits was central to the NRC's responsibilities under the Environmental Policy Act, and confidence that those responsibilities had been faithfully discharged was central to the moral authority of any decision. No doubt responsibility for the investigation was unclear and diffused;

no doubt a full study would take time when economic and political leaders clamored for haste. Since the problem was complex, a thorough analysis might not provide a simple, or even a clear, answer; certainly that answer would have alienated people with an important stake in the process. But among the several failures of governmental process at Seabrook, the performance on alternate sites was outstanding. The whimper with which the curtain fell offered no recompense to anyone for years of frustration.

A balancing of costs and benefits logically required the calculation of both. In the nature of the proceedings, most of the emphasis fell on costs, and the uncertainty of evaluating hazards to the environment and other social costs led to varied estimates and a great deal of argument. The benefits of operating Seabrook Station, by contrast, were relatively easily foreseen: some incidental economic stimulation through purchases, taxes, and employment, and the major boon of an adequate supply of electricity to meet rising demand. To the utilities and their supporters, and sometimes to the several regulatory agencies, that beneficial result seemed worth a very high cost indeed. One of the NRC's appeal boards, for instance, wrote in another case in 1978 that "genuinely needed electricity can be perceived as 'priceless.'" If that appeared to overstate the case slightly, the board continued, at least the value of the plant's output would increase in proportion to costs of production.

> Once it has been determined that a generating facility is needed to meet real demand, . . . the final cost-benefit balance will almost always favor the plant, simply because the benefit of meeting real demand is enormous—and the adverse consequences of not meeting that demand are serious.[45]

The attitude behind that opinion lent force to the intervenors' constant allegation that they faced an impossible task. In estimating a need for power, utilities had most of the technical advantages and all of the responsibility, a situation that ordinarily evoked sympathetic treatment from regulatory agencies. Further, at least by implication, the choice of fuel to be used to generate the "priceless" current necessary to meet "real demand" rested with those who would own and run the plant. Operation of a utility, after all, is management's task, not that of public boards, and, by extension, none of the public's business. To challenge that logic and, from the viewpoint of intervenors, its unfortunate consequence required establishing the converse: No power was needed; demand was not "real." Or, if those negative propositions could not be demonstrated, opponents had to prove that some other fuel was so superior as to require interference with management's prerogative. The difficulty of making either case explains the intervenors' use of the licensing process to force an upward revision of costs, in the hope that at some point the NRC might decide they outweighed the benefits.

The dimension of the hurdle was evident from the outset. Unconvinced by PSNH's forecast of demand, the Public Utilities Commission (PUC) of New Hampshire had checked the figures against those of the regional power pool. The two estimates differed in detail, but the PUC decided the details did not total a substantive difference and concluded that there was indeed need for new generating capacity. Intervenors attacked this judgment before the state's siting committee but lacked the experience and authority that might have persuaded the committee to discard the utility's estimate. Intervenors' witnesses, the committee noted in 1973, merited "little consideration" when in conflict with "some forty-five years of actual performance by the company in successfully projecting and matching generation capabilities to actual needs."[46]

In fact, the company's projections were exaggerated. PSNH and the New England Power Pool assumed a continuing growth in the consumption of electricity regardless of price. As early as 1974, when energy prices jumped as a result of the oil boycott, intervenors pointed to evidence in the utility's own figures of price sensitivity and reduced consumption. One or two years, PSNH argued, did not make a trend and did not require modification of assumptions based on extended experience. In comments in 1974 and 1975, the Federal Power Commission (FPC) agreed that PSNH's forecasting was close to state of the art. Quibbling about assumptions might postpone the need for Seabrook Station by a year or two, but the capacity would be needed soon, the FPC remarked, and the regulatory delay that seemed inevitable might bring supply and demand into fortuitous equilibrium.[47] The contention that Seabrook Station was unnecessary, argued Robert Uhler for the FPC, was not sustainable.

Although federal officials generally accepted PSNH's figures, their rhetoric was less intense than that of the utility. Public Service warned of brownouts, economic recession, and dependence on expensive, imported oil that might dry up at the whim of Arab potentates. Seabrook would save an astronomical amount of oil, thereby avoiding costs for power that would bankrupt business and precipitate widespread unem-

ployment. Any regulatory delay tended to trigger a press release empha-
sizing the company's mounting interest charges, the customers' higher
rates, and the region's stagnating economy. Those same points echoed in
the commission's mail when construction slowed or was temporarily
halted.[48]

Yet there must be an occasional real wolf, as the old fable has it, for
people to credit warning cries. When the economic wolf PSNH her-
alded stayed away from New England's door, the menace began to seem
remote, or perhaps imagined, and those who proclaimed it foolish.
Economic disaster loomed for PSNH itself, but the problem seemed
quite localized and those strident predictions self-serving.

Still PSNH kept them up. In documents issued in 1982, the company
warned that a year's delay would impose the burden of more than a
billion dollars in imported oil, and if the delay lasted as long as three
years (which of course it did), the additional oil bill would exceed $4
billion. The calculation was based on oil at an average price of more
than fifty-eight dollars per barrel, and the actual cost turned out to be less
than a third of that amount. Since PSNH had also overestimated the
amount of electricity needed to meet demand, and therefore the amount
of fuel required for generation, the frightening forecast bore essentially
no relationship to reality.

Predicting the future price of oil was a notoriously difficult task in the
1980s and was not, after all, PSNH's major line of corporate work. But
that sort of plausible excuse did not apply to predicting the demand for
electricity. Early in the 1970s, when Seabrook Station was first proposed, the company had expected demand to increase roughly two and a
half times by the end of the decade, when the plant was to become
productive. In fact, demand increased about 50 percent, and the company was slow to revise its projections. Even after the energy crisis jolted
the complacent notion that demand was unrelated to price, year-to-year
forecasts tended to overstate need by about 7 percent, an error longer
projections compounded. PSNH claimed in 1982 to have improved the
technique by reducing reliance on computerized regression analysis and
giving more emphasis to management's own sales forecast. Because of
the "economic and other life style advantages that New Hampshire has
offered in the past," PSNH continued to use a growth factor larger than
that used in the rest of the nation.[49]

The company had had no lack of advice that its estimates were off the
mark. In 1974 Hendrick Houthakker, a Harvard economist, noted that

PSNH's projections were the only ones he could find that justified construction. No one except the utilities, he wrote, believed that the demand for electricity bore no relationship to price—that demand was "inelastic"—and most authorities believed the converse. An economist from Holy Cross agreed, calling PSNH's forecasts "nothing more than a naive extrapolation of an historical trend," which would result in building unnecessary generating capacity that ratepayers would have to pay for. William Gillen, an independent consultant, called attention to extensive econometric research that demonstrated a direct relationship between price and demand:

> The assumption in the applicants' projection, on the other hand . . . is clearly . . . an . . . error which need not wait until the late 1970s to be tested. Whatever our question about the price elasticity of demand, the one thing we may confidently assume now is that it is not zero. [50]

PSNH's tables, of course, assumed that the relationship was zero.

In addition to an attack on the applicants' statistics, intervenors tried to show that conservation or the use of other fuels could supply any increased demand. By its nature, the case was hypothetical and therefore difficult to make convincingly. Savings from computerized controls and insulation, for instance, were demonstrable, but the total expected reduction in demand for kilowatts was a number that quantified at least as much uncertainty as the utilities' estimates of need for power. Over Dignan's objections, Robert Backus introduced testimony from expert witnesses who, on the basis of investigation for Vermont, were persuaded that wood was both cheaper and safer than uranium as a source of electricity. Other witnesses described heat pumps and solar cells and restrictive shower heads without convincing the licensing board that such devices would be an effective substitute for a pair of nuclear reactors.

Donald Stever thought he might have more persuasive, quantitative evidence after his consultants finished checking PSNH's econometric models in 1975. He asked the ASLB to postpone a decision until his volunteers completed their work. Dignan opposed any delay, noting that the state of New Hampshire, which employed Stever, had already agreed Seabrook's production was needed. The chairman of the licensing board, accustomed to staffing in less penurious polities, simply could not believe that a state's attorney general relied on unpaid university students to carry on the public's research. And John Frysiak was not about to wait for them to complete the assignment; the research went undone, and the board uncritically accepted projections that were far from accurate.[51]

By 1982 statistical accuracy no longer mattered. Although intervenors repeatedly sought reconsideration, the NRC ruled that all questions about the need for power must be resolved in hearings during the construction permit phase. Even "if subsequent facts make the findings seem inappropriate," the commission decided those facts ought to be ignored. The new rule codified the Seabrook experience, where the company's predicted demand for power had only once been close to the "facts" of actual electrical use.

That sort of managerial ineptitude hardly inspired public trust; the same corporate team, after all, had disastrously underestimated the cost of the project. When accidents and public safety became the subject of discussion, there was little reservoir of public confidence for management to draw on. Failure accurately to forecast sales, revenues, and expenses hardly argued that executives would be reliable operators of a potentially hazardous power station. Further, as a result of diminished demand, there appeared by some calculations no need for Seabrook Station to meet the region's need for power. In that case, construction was the ultimate managerial blunder.

Approval of continuing construction at Seabrook indicated that the NRC rated the staff's cost/benefit analysis more favorably than did the Holy Cross professor who called it unworthy of "a passing grade . . . in an undergraduate course in economics." After some quantitative window dressing, he wrote, the conclusion boiled down to an affirmation that "the staff believes that the benefits . . . will outweigh the costs." The commission, in the end, believed that too.[52]

Yet the license for a nuclear plant ought to be based on evidence, not on faith. And the staff, whose task it is to verify the evidence and test the facts, did not have its most shining hour during the Seabrook litigation; the revealing discussion of transmission corridors was only the worst of many bad moments. The utilities also erred badly in estimating costs and predicting demand, which are precisely the tasks management is paid to perform correctly. And the intervenors made tactical mistakes, such as the failure effectively to counter early the case that New England needed more power. The licensing process worked to protect the marsh and to reduce predictable harm to the marine environment. Beyond that, as virtually everyone observed, it produced irritation, escalating costs, and belated and grudging approval for a scheme many thought flawed.

The process was also ultimately incoherent. However logical it was to deal sequentially with the impact of Seabrook Station on this or that aspect of the surrounding environment, the pieces never came together in a way that permitted consideration of the project as a whole. In comparing alternate sites, for example, the question of whether any plant should be built anywhere was evaded. The selection of nuclear fuel, which occasioned the entire controversy, received relatively little direct attention. And the cost/benefit analysis, which ought to have been a synthesis, was rigged from the outset by the almost unexamined proposition that needed power was worth any cost. Periodically, Thomas Dignan complained that the fragmented, disjointed, regulatory process worked a very expensive hardship on his client. The complaint was legitimate and could with justice have been made on behalf of the intervenors and the public at large. In that one respect, the NRC was impartial.

3 | The Opposition

"No, No, No."

Congressman Sterling Cole, slightly bewildered, peered at the crowd and the television cameras. He knew the atomic energy legislation that he and Senator Bourke Hickenlooper had introduced was important, but hearings of the Joint Committee on Atomic Energy had not previously attracted interest of the dimension he encountered that morning in the Capitol. Cole soon realized that he had blundered into the confrontation between Senator Joseph McCarthy and the United States Army, the Washington obsession of the spring of 1954. Redirected, he went off to his own committee room, musing at the paradoxical lack of public interest in the crucial legislative business in which he was absorbed.[1]

He was right about the significance of what became the Atomic Energy Act of 1954. Developed to give practical expression to President Dwight Eisenhower's vision of the peaceful use of atomic fission, the law ended the government's monopoly of atomic technology and laid the basis for private commercial applications of wartime atomic research. There was, in 1954, little national guilt about the use of nuclear weapons, and the prospect that the atom would soon produce power, profits, and abundance for a peaceful world helped to allay any emerging ethical unease. Lewis Strauss, the chairman of the Atomic Energy Commission (AEC), not only promised electricity "too cheap to meter" but attributed knowledge of atomic structure to divine intervention: "A Higher Intelligence decided that man was ready to receive" nature's secret, Strauss wrote; not to use the gift would obstruct God's purpose.[2]

Although the Atomic Energy Act encountered some congressional opposition, the general proposition that fission should be adapted to peaceful use, including the generation of electricity, was not seriously challenged in 1954, nor has it been legislatively reexamined since. Plans for specific nuclear plants have stirred vigorous protests, and nuclear power has become locally controversial. But particular conflicts have had only faint echoes in Congress, where there has been no important effort to reconsider the decision, taken in 1954, to encourage the nuclear industry as a matter of national policy. Congress has replaced the Atomic Energy Commission with the Nuclear Regulatory Commission (NRC), tried to dispose of nuclear waste, and otherwise tinkered with details without facing squarely questions opponents ask: Should nuclear fission be used to generate electricity for profit, and do existing technology and regulations adequately protect the public's health and safety?

Debate in 1954 illustrated evasion of such matters. Senator McCarthy warned that the end of atomic secrecy might endanger national security, but he was no longer a match for a popular president preaching "Atoms for Peace." Labor unions, defenders of the Tennessee Valley Authority and other advocates of public power, and an embattled rear guard of New Dealers argued that atomic energy ought to remain a governmental monopoly; developed with public investment, the atom ought not to generate private profit. But the essential safety of the technology was simply assumed, and the legislative requirement that public health and safety be protected was not detailed and seemed almost an afterthought. The faith that experts who had harnessed the atom's power could be entrusted to insure its safe use was so widely shared that it hardly needed stating.

That faith, however, was not sufficient for insurance companies, which declined to provide liability coverage for private utilities and their major suppliers. The technology was untried and the risks uncertain; without insurance, the opportunity the Atomic Energy Act opened could not be grasped. In 1957, with only minor dissent and little discussion, Congress limited the liability in any nuclear accident to $560 million, which a formula divided among utilities, their insurers, and the federal government. This Price–Anderson Act also guaranteed public access to licensing hearings the AEC was required to hold before permitting construction or operation of a nuclear enterprise. The agency accepted public participation, which it had previously opposed, in order to secure liability protection.

At the outset, the AEC's fear that an uninformed public would clutter and delay licensing proceedings was unwarranted. Only three of the twenty-six applications before 1967 were contested and only twenty-seven of the hundred filed in the decade ending in 1971. The industry benefited from the reflected prestige of atomic scientists, the widespread respect for technical expertise, and the silence of governmental agencies about some of the dangers inherent in the new technology. Believing the subject too complex for laymen, the public simply trusted experts and the government to provide both the benefits of nuclear power and protection from its risks. Orders for commercial reactors, both in the United States and abroad, proliferated in the 1960s. The oil shortage, exacerbated by the Arab boycott in 1973, imparted a special urgency to the effort of electric utilities, especially in the oil-dependent Northeast, to increase nuclear generating capacity. [3]

Popular willingness to try nuclear energy diminished over time. Misgivings became doubts, which then hardened into certainties that neither the industry nor its governmental advocates could dislodge. That evolution was neither quick nor universal and was not entirely the fault of the utilities or the NRC, which replaced the AEC in 1975. Social currents, such as a growing environmental concern, and disillusioning events, such as the government's unresponsiveness to popular antiwar protest, created an unhealthy climate for governmental sponsors of a new, and admittedly hazardous, industrial process. A recurrence of egalitarian populism, which achieved varied expression in a pervasive challenge to social, cultural, and political authority, led not only to a disparaging of racial barriers and traditional standards of conduct but also to doubts about technical expertise, such as that of nuclear engineers. The context, in other words, of Woodstock, My Lai, Watergate, and Watts and an inability to explain, let alone control, ruinous inflation moderated the reflexive enthusiasm that had frequently greeted plans for nuclear generating stations elsewhere when the Public Service Company of New Hampshire (PSNH) tried to promote a plant on the New Hampshire seacoast at Seabrook.

Although the timing was not propitious, the task did not seem insuperable. A survey of public opinion that PSNH ordered in 1972 showed that New Hampshire residents approved of nuclear fuel as an environmentally responsible means of meeting a perceived regional need for energy. Should a new generating facility be required, more than 50 percent of respondents favored nuclear fuel, and a plurality of those with opinions endorsed the Seabrook proposal.

But the survey was no mandate. When asked whether they preferred economic growth or environmental protection, an overwhelming 70 percent of those polled chose environmental protection, and only 11 percent thought the existing rate of growth too slow. Other data in the survey confirmed this sensitivity to the environment, which was particularly acute among seacoast residents and younger respondents. The public apparently feared thermal pollution of the ocean and damage to the marsh more than radiation or other threats to health. PSNH could, therefore, infer public confidence in the technology and predict substantial support for the plant if environmental harm were seen to be eliminated or substantially reduced.[4]

Perhaps the company's public relations campaign, which stressed safety, had reduced apprehension on that score. Advertisements showed a landscaped "nuclear park," where families enjoyed picnics amid flowers, shrubs, and cooling towers. As a result of expensive federal research, nuclear plants were safe, clean, efficient, and "compatible with the environment," the advertisement proclaimed, and those who lived nearby were exposed to less radiation than the atmospheric dosage everyone received. (How the plant actually reduced ambient radiation was not explained.) Electricity seemed but one important amenity that would result from the attractive addition PSNH wanted to make to the neighborhood.[5]

The Seacoast Anti-Pollution League (SAPL) was not persuaded. Begun in 1969 to combat the abortive first scheme for a nuclear plant at Seabrook, the organization used the months after that project was deferred to promote environmental sensitivity and extend the membership. Never large—the core of the organization in the early days consisted of a dozen people or fewer—and always short of money, SAPL depended on the energy and resources of dedicated volunteers to fend off the utility and its political and financial allies. The odds seemed distinctly uneven.

At the outset, SAPL subordinated opposition to nuclear power to defense of the coastal environment, though the two objectives were linked in the organization's first statement of purpose: "The Seacoast Anti-Pollution League is . . . organized to work toward a deferral of the proposed nuclear plant at Seabrook" and is vigorously opposed "to any alteration or degradation of New Hampshire's tidal marshes or estuaries." In their appeal for public support, SAPL's founders portrayed their organization as moderate and constructive, committed to education, reason, and legal process; they seemed anxious not to appear doctrinaire foes of economic and industrial growth. Some members did not object to construction of a nuclear plant, provided adequate safeguards for the coastal environment could be assured. Radiation, to this faction, seemed no more imminent a hazard than phosphates or other

chemicals, sewage, greed, or ignorance. New Hampshire's state legislature was unresponsive to amateur lobbying for environmental protection, however, an attitude Walter Tingle, an airline pilot and SAPL's
first president, related to the state's abiding opposition to taxation.

> And the New Hampshireites rose up with a great hue and cry and said . . .
> "If we are taxed we will not be able to afford a second snowmobile, nor a new
> trailbike. . . . And the boat with the motor of many horses . . . will vanish
> away, and the smaller boat we now own will seem abominable to us."

"The people were taxed not," Tingle's fable continued, and real estate
developers, unchecked by law, filled the wetlands, polluted the streams,
and covered the terrain with asphalt and fences. The "New Hamp-
shireites" discovered that "even their boats were no fun because of the
stink of the waters, of which it is said they could practically be walked
upon." For lack of intelligent environmental legislation, residents were
reduced to prayer and the hope of eventual natural restoration. "And
they will wait a long time—even until hell freezeth over."[6]

Tingle made the same appeal less whimsically to the state's outdoor
sports enthusiasts. He noted that *Field and Stream* had commended
SAPL's legal efforts and asked for the financial support of "sportsmen . . . who care what is happening to prime fish and game habitat."
Those who enjoyed hunting, fishing, and clamming, Tingle wrote,
should help block plans for the nuclear plant at Seabrook or else
they would have themselves "to blame when the last duck marsh is
paved . . . , the last river polluted," and "the last clam flat dredged
away."

Tingle's writing reflected his organization's determination to attract
and retain support beyond those who were ideologically committed to
the fight against nuclear power. SAPL thwarted local developers, who
aspired to fill wetlands, and helped persuade a state agency to drop plans
for a housing project on the marsh. The league's periodic column in a
local weekly pointed to the ecological importance of the estuary, which
had often been disparaged as a source of mosquitoes, foul odors, and
other nuisances; those columns undoubtedly had some responsibility for
the heightened environmental consciousness revealed in PSNH's poll.
But the nuclear plant seemed the central threat to the marsh, and PSNH
consequently SAPL's major foe; opposition was unyielding from the
moment in 1972 when the utility revived the project and asked approval
from the state siting agency. "It is clear," noted the SAPL newsletter, that
the utility's managers "have not yet fathomed the depth of our resolve to

oppose this project." The officers hired a lawyer, filed a brief with the state siting committee, and worried about raising the $21,000 their ambitious budget required; a family membership cost five dollars at the time, and there were perhaps 400 enrolled members.[7]

The organization did not enroll every skeptic. A reporter for the *Boston Globe* talked with lobstermen who sailed out of Hampton Harbor, some of whom by their own admission had paid little attention to PSNH's plans for their home port. But if the company was "going to harm this river or the ocean," they knew where they stood without deep thought: "to hell with it." At least one seasoned lobsterman had considered the proposed plant at length, and Irving Jones dismissed the environmental reassurance of those on the PSNH payroll. "I imagine they got their learning out of a book somewhere," Jones said. The ocean, as those with long acquaintance knew it, was too dynamic for quick observation: "you must take at least a 25-year period to even begin to know somewhere near what's likely to happen if you introduce major new factors into the area." Life in the water tended to run in cycles: some years pollock, some years smelt, and other years clams. Data for any particular year, no matter what they seemed to prove, did not convince Irving Jones. Bruce Brown, another lobsterman and an elected selectman of Seabrook, wondered why Jones and all the other self-styled experts on the docks had not yet received appointments to the Atomic Energy Commission if they were so wise. As for Brown himself, if the AEC said the plant could be built, that was enough for him.

That sort of faith in the regulatory agency helped PSNH make the case for the plant; governmental oversight would assure safety, both to persons and to the environment. The utility simply contradicted assertions that harm might befall the marsh or ocean and maintained instead that Seabrook was "the best site which could be selected anywhere from Virginia to Northern Maine." Far from interfering with the recreational industry and the beaches, the plant would attract tourists and enhance the region's natural attractions.[8]

Contradiction and reassurance rested on the unstated premise that a multimillion-dollar corporation and its allies in banks, law firms, and political bureaucracies had better information than a few disgruntled lobstermen and naive, underfunded volunteers. Since the industry had, in effect, a monopoly on nuclear wisdom, and since that wisdom encouraged continued expansion, objections were never well founded and derived at best from misperception and at worst from deliberate perversity. Misinformation could be corrected through public education; obstruction would simply be overcome by the industry's economic and political power. That analysis suggested two tactical approaches, which the nuclear industry tended to pursue simultaneously: a so-called informational effort to shape public perception, and a regulatory process that in the final analysis approved the industry's plans.

The strategy evolved early. The industry characterized intervenors as "uninformed people, who needed to be taught how risk-free and beneficial nuclear power . . . was," a study prepared for the Rand Corporation noted. Rather than confront substantive concerns, regulators attempted "to change public attitudes through better information programs and cosmetic tricks." An agency document of 1965 suggested that open hearings ought "to give the public a first-hand impression of the applicant's character and competence" and to demonstrate the AEC's dedication to safety. A writer for the *Notre Dame Lawyer* pierced that facade: Public hearings were instead "mere window dressing" for decisions made by the regulatory staff "behind closed doors and away from public scrutiny."[9]

Seabrook began to look like a case in point. Early staff qualms about the site, about emergency planning, and about the beach population appeared to lose their urgency when members of the public joined the discussion. Once the application was formally submitted, intervenors considered the utility and the regulatory staff informal allies and felt that the government represented the corporation rather than the public. They feared a result that was rigged rather than reasoned, and they suspected that much of what they believed to be persuasive evidence and argument went unheard, because it stemmed from a view of the world that nuclear advocates rejected out of hand.

"Why am I here tonight?" Dr. Frank Graf asked the Atomic Safety and Licensing Board in 1986. A decade before, the surgeon said, he had appeared before another "tired, bored, anesthetized, salaried" panel considering the Seabrook plant and said the same things. Events since, at Three Mile Island and Chernobyl, had unfortunately borne out assertions he had made then about the risks inherent in nuclear technology. He hoped he would not have to repeat himself many more times because, Dr. Graf confessed, "I am bored by my own testimony," which he and everyone else had heard before. He hoped the board's anesthesia would soon wear off and that he would be heard and heeded.[10] An attorney for intervenors in another hearing had a similar reaction but used a different figure of speech: "I really felt like a Jew in Germany," he said. The judges "had their minds made up," and nothing he could do reached them. "I tried joking with them; I tried yelling at them. And they were nice people. . . . It was just like the reality in my head and the reality in their heads were just extraordinarily different."[11]

That was one way of remarking that minds on both sides of the issue had closed, a conclusion with abundant illustration. Both the presiding judge at Seabrook hearings and a lawyer for the Department of Energy told separate, and hostile, New Hampshire audiences that their fear of nuclear technology stemmed from ignorance that authoritative governmental information would repair. In both cases, opponents responded that it was their knowledge of nuclear power, not the absence of knowledge, that led to fear. On any topic, from alternative sources of electrical power to emergency planning, from environmental pollution to the number of people at risk on the beach, there were multiple sets of mutually contradictory facts as well as multiple interpretations. And the realities those facts purport to describe do indeed differ markedly from one another.[12]

The New Hampshire legislation establishing a committee to site future electrical generating stations provided for a counsel for the public, a post filled by Assistant Attorney General Donald Stever, to assure adequate popular representation in adjudicatory hearings. The Audubon Society and the Society for the Protection of New Hampshire Forests (SPNHF), two traditional conservation organizations that owned land the Seabrook project might affect, decided in 1972 to guard their interests during the siting committee's deliberations; jointly, the organizations hired Robert Backus, a young Manchester attorney, to represent them. The Seacoast Anti-Pollution League secured the legal services of Berlin, Roisman, and Kessler, a Washington, D.C., law firm with experience in contesting the Vermont Yankee nuclear plant. Public Service of New Hampshire added Thomas Dignan, of the Boston firm of Ropes and Gray, to its corporate counsel. For many of those involved, in spite of their effort, the case became a career.

Several of the issues were apparent at the outset. SAPL argued from the beginning that PSNH's load forecasts overestimated the region's future demand for electrical energy. A nuclear plant, in any case, was the wrong way to supply whatever need might evolve, because the technology was both unsafe and unreliable, SAPL maintained. Construction and operation of a plant at Seabrook would unacceptably alter the coastal environment; in particular, the proposed cooling system, which at first involved conducting seawater through a ditch in the marsh, would ruin the natural balance in the estuary and in the ocean beyond. Finally, SAPL pointed out, PSNH's own studies demonstrated the superiority of other sites. In fifteen years of subsequent legal wrangling, the contestants found other topics to debate, such as emergency planning and the company's financial stability. But the agenda before the state siting committee sufficed for some time.

It was more than enough for the committee itself. Because the Seabrook application provided the first test of the new siting law, the committee had no precedent; because the legislature provided no budget, the committee had neither counsel nor staff. The hearings, therefore, which dragged on for more than a year, are no model of adjudicatory efficiency or fairness. In the middle of the process, furthermore, the committee became the focus of unconcealed political pressure when a new governor took office. Walter Peterson, who left office in January 1973, had allowed the committee to find its own way; Meldrim Thomson, who replaced him, trumpeted his contempt for environmentalists, denounced delay and due process, and impatiently insisted that all obstacles to construction be removed, including obstacles on the state's payroll, some of whom held membership on the siting committee.

Members of the committee represented state agencies charged with environmental protection, and PSNH submitted data to demonstrate that the plant and its cooling system would have essentially no effect on marine life. John Clark, an oceanographic consultant engaged by SAPL's attorneys, testified that PSNH's own data, whose accuracy he did not concede, demonstrated conclusively that the cooling system alone would irreparably damage the coastal environment. As the hearings concluded and the committee commenced deliberation, PSNH substituted tunnels for the environmentally disastrous canal in the marsh, a concession that permitted the group to ignore months of testimony and finesse conflicting evidence. Although the committee had shown little urgency during more than a year of desultory proceedings, it pleaded a tight schedule and refused to allow SAPL to question utility executives about this fundamental design change. Assessment by Clark or other independent experts was out of the question.

Bereft of the central portion of her case, Diane Curran, SAPL's attorney, developed the argument that the region had no need for additional generating capacity. Her summation noted inconsistencies in PSNH's predictions of future load and a complete failure to account for the effect of rising energy prices. She appealed to New Hampshire's famous parochialism with the claim that the plant would serve the needs of other New England states more than those of New Hampshire. PSNH, she implied, had been hoodwinked by more sophisticated partners into assuming financial and regulatory burdens they did not want. In particular, utilities in southern New England refused to expend valuable sites for future plants to build presently unnecessary generating capacity. [13]

The argument had the ingenuity of desperation, and it availed SAPL nothing. In the summer of 1973, the committee issued a one-page decision that accepted PSNH's case and approved the site. Disappointed intervenors could expect neither solace nor assistance from the state's executive branch, for Governor Thomson had openly lobbied the siting committee and tried some crude arm twisting on the federal bureaucracy as well. But they hoped for more evenhanded treatment from the state's judiciary, and they hired Robert Backus to appeal the siting decision to the New Hampshire Supreme Court. The court agreed that the committee's perfunctory opinion was defective and ordered a more detailed explanation. In 1975, the siting committee provided an amended version that Donald Stever branded "little better than its predecessor." But intervenors, preoccupied with federal proceedings and financially strapped, chose not to renew the appeal. [14]

SAPL's financial crisis was a perpetual fact of organizational life. The group's survival rested on the selfless energy of a few volunteers who served as quasi paralegals, raised money, wrote press releases, and kept one another's spirits up in the face of countless disappointments. They were ordinary citizens—housewives, carpenters, pilots, commercial fishermen, and retired folk—and they had an outsize faith both in legal process and in their own convictions.

But they had, in the beginning, no idea what those beliefs would cost, though the officers knew in 1973 that expenses in connection with the state hearings had certainly exceeded estimates and exhausted the treasury. In August the board voted to seek negotiated settlements with John Clark and Berlin, Roisman, and Kessler, because the discouraging result of the siting hearings had abruptly reduced the flow of funds. "The business people at the beach who [had] promised help" had reneged, SAPL's president explained to Clark, and public support had "dwindled to a few hundred dollars per month or less." Under the circumstances, paying Clark's bill, which was $3,000 more than the league had authorized, was impossible. It required a year, hundreds of small gifts, and a very successful craft fair to pay debts accrued during the state hearings.

The outlook, in the board's view, did not warrant throwing scarce money at further appeals. [15]

But SAPL's public optimism never wavered. "Even though we got several adverse decisions from various state commissions," the SAPL *News* noted early in 1974, "our efforts have prepared the groundwork for the case before the AEC." The organization expected, "at the very least . . . more fair and equal treatment from the AEC than we received from our own state officials." It was optimistic, and not very accurate, prophecy. [16]

The brave public front masked an organizational transition. Many of SAPL's early leaders, worn out after five years of unremitting and apparently unrewarded effort, saw the league through the budgetary crisis of 1973–4, for which they felt responsible, and then sharply reduced their effort. New leadership proved less effective and less dedicated, on the whole, than the founding group had been. By 1975, SAPL was sinking. In effect, Robert Backus, who served as SAPL's counsel through years of litigation, provided continuity and unpaid leadership, as well as his legal service, for which he was usually paid tardily.

SAPL's unlikely new president in 1975 was Guy Chichester, a self-employed builder who had served an agitational apprenticeship in the effort to block a coastal oil refinery. Chichester's election was a symptom of organizational malaise, for, by his own admission, he had known nothing of nuclear power before 1975 and had had no important association with the league. His predecessor could not find volunteers to serve on the nominating committee that eventually asked Chichester to serve, and he was not that committee's first choice. The SAPL *News* rationalized this disarray by noting that the technical expertise required of officers during the early stages of the struggle over Seabrook Station would not be needed now that litigation was drawing to an end. New issues would require new vision. [17]

Chichester's vision was indeed new, though it took more than a year for old-line SAPL members to realize just how novel, and unacceptable, it was. The new president thought the organization too middle-class, too committed to legal process, too politically unsophisticated, and, in its concentration on the environment, focused on the wrong issue. Chichester believed that legal action without political intervention could not halt construction of Seabrook Station and that environmental protection roused less political passion than would fear of nuclear power. His politics were not partisan or even electoral; what he had in mind might better be described as a political consciousness to complement the environmental consciousness that had energized SAPL in the first place. The licensing process, in Chichester's view, on both state and federal levels, was only one manifestation of a political system arranged to serve elites and to deceive ordinary citizens. Unquestioning acceptance of governmental rules, therefore, led to defeat and impotence, a future he saw for SAPL unless the organization stimulated broad political opposition to the nuclear industry and the governments that protected it in the guise of regulation.

So Chichester began to use SAPL's rostrum to make political statements. He hoped to goad Governor Thomson into foolish rejoinders, and he hoped to encourage and attract other antinuclear activists. He modified SAPL's rhetorical style, for instance, when he accused the New Hampshire Public Utilities Commission of heeding corporate masters. The three commissioners ought to be replaced, Chichester said, and "put out to pasture along with the other old bulls . . . who don't know the difference between rape and service." He urged people to attend licensing hearings that took place in the summer of 1975, and he called radio stations and reporters to point out incidents revealing the undemocratic nature of the proceedings. He sought out people from the antinuclear movement when they visited the seacoast that summer, and he kept in touch by telephone when they left. His enormous telephone bill and his general budgetary indifference precipitated a heated discussion in March 1976, during which the SAPL board decided against demanding his immediate resignation. Dorothy Anderson, SAPL's vice-president, backed Chichester in the meeting, but a candid, subsequent letter outlined the limit of her support.

The board's irritation, Anderson wrote, stemmed not only from Chichester's failure to account for expenses but also from his manifest disregard for their concern about the matter. "Being president of an organization," Anderson chided, "does not give you the right to do as you please." He had imposed on people without evident appreciation, made demands of people who ought to have been asked, and appeared to enjoy "throwing [his] weight around." His piece for the newsletter was always late and never redeemed by either effort or quality; indeed, Anderson said, "I'm really embarrassed by that part of the newsletter." In effect, she charged, Chichester was lazy, and she believed the league would function more smoothly "if you were willing to make the same demands on yourself that you make on other people."[18]

Even in a private letter, Anderson emphasized money and leadership style and left unmentioned tactical and ideological differences that were at least as basic. The board probably did not want to advertise internal division that might inhibit fund raising when SAPL's debt exceeded $10,000 and appeal of the decision authorizing construction would require at least another $5,000. But Holly Meiklejohn, a volunteer who had herself raised thousands of dollars for the organization and who rarely used her considerable influence to shape policy, left no doubt about where she stood. In July 1976, shortly after the Atomic Safety and Licensing Board authorized construction, she urged SAPL to keep to its traditional legal strategy and explicitly to disavow any extralegal response to construction. "To say that the only way to stop the nuclear plant is to 'pitch our tents in front of the bulldozers' is not true," she wrote. Readers of her forceful letter knew that Guy Chichester had talked more than any other member of the league about obstruction and civil disobedience.[19]

By the time Meiklejohn's letter was read, her reference needed no explication: On August 1, 600 protesters attempted to seize the construction site; 18 were arrested. They were the vanguard of the Clamshell Alliance, of which Chichester appeared to be the spokesman; his association with SAPL had ended. Three weeks later, police arrested another 180 demonstrators. An October event attracted more than 200. With some exaggeration, an article in *Nation* called attention to this "Nuclear War by the Sea." While SAPL went forward in the courts, Chichester and the Clams adapted civil disobedience and symbolic direct action, as taught by the American Friends Service Committee, practiced by the civil rights movement, and exemplified by antinuclear activists abroad, to circumstances on the New Hampshire seacoast. Cells, called affinity groups, gathered throughout New England to develop mutual trust, share ideas, devise tactics, and make decisions; the combination of these groups—the Clamshell Alliance—was a diffuse organization, committed to observe democratic principles and to oppose economic power and centralized control over energy distribution and the lives of individuals. The alliance set policies after endless discussion—often requiring months—had produced agreement by consensus or, perhaps, exhaustion: "we are fighting corporate power," one organizer wrote, "not imitating it."[20]

The Clams displayed the style and idealism popularly associated with the Woodstock generation. They wore jeans and headbands and carried guitars and backpacks; their songs often seemed more articulate than their speech, their actions more eloquent than song. Clams espoused tactics more than ideology; they reacted rather than proposed. They displayed the outward manifestations of earlier social movements against racial discrimination and war, and they had a similar internal certainty. They had the energy of zealots and a dedication to their cause that may have substituted for children or jobs or property that many of them had not yet acquired. They were easily and frequently derided as young and inexperienced and outside most behavioral and political norms—characteristics that explained the brevity, and much of the intensity, of their passage through the Seabrook epic.

For two years, as legal processes moved undramatically forward and construction began, halted, and began again, the Clams personified a colorful, personal, and continuing hostility to nuclear power. They refused to concede the legitimacy of federal decrees or the inevitability of further construction; they found the NRC and rolling cement trucks equally unpersuasive. In spite of the steadfast support labor unions gave the project, Clams distributed leaflets at the site explaining that they sympathized with workers who needed jobs but nevertheless opposed their employer. Fortunately, Clam leaflets maintained, workers need not choose between jobs and opposition to nuclear power; "we demand secure jobs for all displaced workers in the renewable energy field," which ought to employ "more than twice" the number of people at work building nuclear plants. Workers, earning the best wages most of them had ever received, were notably unmoved. [21]

What the Clams called "leafleting" was not the main event, and the construction trades were not the primary audience. Civil disobedience, begun in 1976, continued with a much larger demonstration at the end of April 1977. Clams made no secret of their intent to occupy the site, their hope thereby to stall construction, their fear that state authorities might use violence, and their belief that such manifestations of conscience might move the public to enlist in the antinuclear crusade. The organization conducted workshops in nonviolent response to police and forbade fence cutting, property destruction, drugs, alcohol, and all illegal acts except trespass. Members told the Seabrook police of their plans, secured campsites from nearby sympathizers, and presumed that Governor Thomson would overreact.

He obliged. Although Governor Michael Dukakis decided the event posed no serious threat to order and declined to lend officers from Massachusetts, Thomson mustered busloads of New Hampshire police and deployed them to arrest more than 1,400 demonstrators. Arraigned at the Portsmouth Armory, most of those arrested believed they would be released without bail for a later trial. But, apparently at Thomson's insistence, court authorities began requiring bail of $1,500, which the prisoners uniformly declined to post, a decision that forced the state to hold them for trial.

Governor Thomson's central claim to political popularity was his pledge to keep New Hampshire the nation's only state with no tax on either income or sales. Thus the state had a tight budget and restricted state services, on both of which 1,400 unexpected involuntary guests placed considerable strain. Nervous local officials maneuvered to deflect costs for the care and feeding of prisoners who spent two weeks in temporary quarters in the state's armories. Thomson whined that this hospitality was costing New Hampshire more than $50,000 each day, and he asked for help from the federal government, the nuclear industry, and others interested in maintaining law and order. On behalf of the prisoners, the American Civil Liberties Union sued the state for damages, an action that a federal judge promptly dismissed. A county attorney finally negotiated a mass release without bond pending a trial that was so long delayed that charges had to be dropped. The state emerged from the contretemps poorer and without much dignity.[22]

Fourteen hundred arrests captured the nation's attention, as the Clamshell Alliance had intended. But civil liberties issues diluted the alliance's statement about nuclear power, and the demonstration failed to attract the broader social base that Clams sought. Rather than rethink the strategy, however, they planned a bigger and better event for 1978, even though the context in which they planned had changed. The larking irresponsibility of 1977 became troubled paranoia a year later. A few individuals, not covered in the agreement with the county attorney, received unexpected six-month sentences for criminal trespass, though the prosecutor had asked only for fifteen days. The minutes of a Clam committee recorded one reaction: The authorities had prosecuted selectively, sentenced severely, and scared "the shit out of everybody else, along with confusing us in the process."[23] Nor were long jail terms the only instances of official intimidation: State police questioned seacoast sympathizers; city police kept the alliance's office under intermittent surveillance; telephones seemed to be tapped and mail opened; various officials were using what Chichester called "the rhetoric of violence," with talk of dogs and hoses and gas. Official threats stimulated unofficial mimics, and demonstrators had to develop a tolerance for menace that included games of "chicken" with demonstrating pedestrians, guns displayed in vehicle windows, and other manifestations of would-be vigilante action.

The Clams also suspected that some affinity groups included provocateurs and knew that a few members tattled to state police. Further, other Clams, more radical than most but not necessarily provocateurs, thought the demonstrations of 1977 too timid and disavowed nonviolence. "You can't halt construction," one of them wrote, "without interfering with the workers." Clams ought to arrive in Seabrook in 1978 prepared to cut fences, sabotage construction equipment, block access to the site, resist arrest, and fight the National Guard. Caught between that sort of advocacy and Governor Thomson, the planning group found consensus difficult to reach. [24]

But, bit by bit, consensus emerged, and the script looked much like that of 1977. The handbook reiterated the alliance's commitment to nonviolence, which was spelled out in a code of conduct for the four-day occupation that was to begin on June 24. Participants were urged to bring displays of alternative energy devices to educate spectators, and not to bring small children, though the committee could not quite agree to ban them. The group planning to blockade drilling rigs in the ocean seemed to revive the mellow good fellowship of the year before as they contemplated a summer day at sea. But counsel for the Hudson River environmental group that had thought about sending the sailing ship *Clearwater* to the fete needed something more substantial, such as legal assurance that the ship would not be impounded nor her crew arrested. About 5,000 people signified their intention to come to Seabrook, and the Clams expected two or three times that many. Expenses seemed likely to reach $20,000; the organization had $565 in the bank, but financial responsibility had never been chief among the alliance's virtues.

Perhaps, in fact, the crowd met the prediction; Clams claimed that 20,000 people showed up. But the visitors did not see a demonstration, and there were no arrests. Instead, they saw what a sloppy broadside called an "unprecedented opportunity to bring the question of nuclear power to a large public forum" through exhibits, speeches, and a rally, all of which was presented in cooperation with the state and PSNH. Clams would have preferred, they said, to occupy the site as planned, but other opponents of Seabrook Station had asked them to accept the state's offer of access to a portion of the site, permission to leaflet, and portable toilets.

The decision to talk with the state was not easily taken, and the resulting deal was widely disapproved within the alliance. But Clams who tried to extort better terms from Attorney General Thomas Rath really had nothing to trade. The state's suggested compromise had been well publicized, and it appealed to those on the seacoast whose assistance the alliance needed for campsites and other logistical support. Local politicians, opposed to construction, also opposed extralegal measures to halt it. Clams themselves feared an outbreak of violence and could not be sure their own radical associates would be blameless if it occurred. To mollify those whose conviction demanded expression in civil disobedience, the alliance promised to follow the event at Seabrook with a sit-in at the NRC offices in Washington. That demonstration did not have the urgency of an occupation at Seabrook and drew little attention and not much participation. The settlement may have been necessary, but it robbed the Clams of their distinguishing verve.[25]

And of a fair proportion of the membership as well. Some simply went back to family, school, or job and dropped out of antinuclear agitation. Militant leftists intended to preserve the alliance's distinction by progressing from civil disobedience to direct action. Manifestos from groups in Boston, Brooklyn, and Plainfield, Vermont, pointed out the futility of moral pressure—that is, civil disobedience—and advocated sabotage, obstruction, and arrest instead. Observers might have learned something about solar power at the Seabrook bazaar, Vermont dissidents conceded, but they had received no instruction in techniques that politically aware individuals might employ to change a system that imposed nuclear power. That was the missing dimension in a peaceful demonstration; that had been the Clamshell Alliance's special insight and contribution, which was now obscured. "Ordinary people," the Spruce Mountain group continued, "have the right to reclaim . . . social power which self-styled 'representative' political leaders and elites have usurped from us." The reference was to compromising Clams as well as to governmental officials.[26]

After 1978, though individual Clams acted on their continuing hostility to Seabrook Station, the alliance itself, as a coordinating body, diminished in importance. Demonstrators migrated to Manchester, Washington, Boston, and Wall Street, and always back to Seabrook, with productions of what too often seemed uninspired theater. There was occasional originality and wit, and some dignity on the part of people whose conscience required them to oppose the state. But they were the actions of individuals, not a movement.

In August 1978, for example, after one of the construction interruptions caused by legal maneuvering, workers returned to Seabrook to discover a large antinuclear banner and six people chained to a crane.

One of the faces was familiar; Robert ("Rennie") Cushing, a young Seabrook native and one of the most fervent of the Seacoast Clams, had already logged several arrests. Two others on the crane were Roman Catholic nuns. The banner and the chains were removed, the crane started, and the demonstrators arrested; Cushing turned their trial the following spring into a circus. "Playing the part of Everywoman's lovably rebellious grandson" to a jury of grandmothers, a Boston reporter wrote, Cushing apologized for his lack of legal training and respectfully mocked the legal system. He asked the judge to disqualify himself. Denied. He asked the judge to disqualify the jury. Denied. He asked a witness from PSNH how he would feel if his hometown voted to bar a disruptive plant that was built anyway. Objection sustained. He tried to give jurors a quick course in antinuclear orthodoxy through a series of questions so framed as to require hostile witnesses to provide no answer at all. All routinely ruled out by the judge. Sympathetic Clams, mustered in the yard outside, brought flowers to the defendants; with a flourish, Cushing presented them to the court stenographer. The judge ordered them removed. When the prosecutor concluded, Cushing summarized his case for him: "What you have is six individuals on a crane; what you don't have, your Honor, is a crime." Cushing moved for a directed verdict for the defendants. Denied.

Each defendant addressed the jury about conscience and nonviolence. Cushing called a parade of character witnesses and examined each about the hazards of nuclear power. Foreseeing an endless trial, the prosecutor objected and was invariably sustained; the issue was trespass, not nuclear power. Cushing switched his line of questioning: "Do you believe defendants are entitled to an impartial judge?" The objection was overruled. Cushing pushed on with a question about sentencing, since the presiding judge had previously sentenced Clams to terms longer than prosecutors had requested. Objection sustained. Well, said Cushing innocently to Judge Wayne Mullavey, "May I ask about justice or is there no place for that in this court room?" Mullavey had heard enough. He dictated an order prohibiting further use of the proceeding as a forum for discussion of nuclear power, and he provided the defendants with counsel to warn them about contempt of court. When Cushing persisted, he was sentenced to fifteen days for contempt. Shielded by his codefendants, he refused to surrender when sheriff's deputies attempted to arrest him. They jostled furniture, spectators, and the court's decorum before they secured Cushing and hauled him off to another courtroom to answer the charge. "What we've seen here," said Cushing, "is reminiscent of Alice in Wonderland: first comes the sentence, then the trial, and the crime comes last of all." Mullavey regained his judicial composure, declared a mistrial, and vacated the contempt order. "Wayne said he made a mistake," was Cushing's summary of the two-day fiasco.

But Cushing and friends had also made a mistake if they thought their actions, either at Seabrook or in Mullavey's court, had won converts to the antinuclear cause. Spectators doubted that Cushing had made much headway with the stolid jury, and one reporter asked a thoughtful state policeman about the trial as the defendants tried to make their case. "They're putting nuclear power on trial in there," he replied.

> I think they're hurting themselves with the people of New Hampshire. Lookit, if I stop you for going 80 miles per hour in a 55 mile zone, I don't want to hear you tell me about the efficiency of a six-cylinder engine versus an eight-cylinder. It's the same in there.[27]

The policeman dismissed the antinuclear argument as logically irrelevant to a charge of trespass, which was good Yankee common sense, even if his traffic analogy was not perfect. Nor was he distracted by the Clams' style, which sometimes obscured their critique of American institutions. Supporters of nuclear power, for instance, caricatured the Clams' flamboyance to discredit them and, by association, the entire movement that Clams for a couple of years came to symbolize. George Gilder, whose economic ideas enjoyed a brief vogue among Republicans in 1980, wrote of "vegetarians in leather jackets," who drove "their imported cars to Seabrook listening to the Grateful Dead on their Japanese tape decks amid a marijuana haze." An editorial in the conservative *National Review* ridiculed the arguments of nuclear opponents, who foresaw that

> Terrorists will fabricate bombs in basements; governments, in reply, will abolish democracy; wastes will poison the soil for millions of years; cargoes of toxic fuel will careen recklessly across the republic, while clouds of radioactivity, squirted from casually ruptured reactors, will creep about the landscape like malevolent fog banks.[28]

Mrs. Benjamin Brideau, who lived not far from Seabrook in Rochester, New Hampshire, made her point with less literary pretense. "I want the nuclear plant," she wrote Chairman William Anders of the NRC. "Just because there are a few 'kooks' around here (and most of them come from other states) who want to stop progress so they can have the woods to themselves does not mean we in New Hampshire are all stupid." Another New England supporter of Seabrook Station wondered in a letter to the NRC whether Communists had some responsibility for delayed construction, and a conservative columnist asserted that in fact there was "evidence of Russian involvement," which, "though not abundant," was "conclusive." A nuclear executive thought perhaps "a socialist form of government" was the objective, rather than communism, an observation he based on the movement's hostility to "anything big." In their emphasis on a "sleeping-bag society," nuclear opponents seemed unworldly to those who thought of themselves as practical people. A scientist from Brookhaven National Laboratory, which performed extensive contractual research for the NRC, was bewildered by the whole controversy. Perhaps, he mused, people want a "simpler life," or "to go back to the agrarian concept."[29]

The Heritage Foundation, a conservative Washington think tank, thought the case for nuclear power ought to rest on more accurate information and more rigorous analysis, which Milton Copulos attempted to provide. He knew to the dime the cost to the public of Clam demonstrations in 1977 and exactly how much each PSNH customer would pay for the twenty-three and three-quarters months of delay that bureaucratic snarls and obstruction by intervenors had caused. He calculated that construction workers had lost exactly $132,782,494.70 in wages and the plant's owners a rounded $419 million as a result of suspension of the construction permit in 1978. The numbers rolled on across the pages, with a precision that may have persuaded the Heritage Foundation but which was as deceptive as Clamshell propaganda. Copulos assumed, for example, that wages were lost rather than postponed when construction paused, and he uncritically accepted management's forecasts as the basis for calculations when the history of the project demonstrated management's inability accurately to forecast anything. Delay, similarly, was never the result of weather or deficient engineering or a failure by suppliers, all of which in fact affected the schedule, but always a function of bureaucrats and intervenors, who were familiar objects of scorn at the Heritage Foundation. (Intervenors, of course, found bizarre the notion that they were in any way allied with governmental officials, either by design or by circumstance.)

Bureaucrats had invented a Byzantine and expensive licensing process, Copulos charged with considerable accuracy, that enabled intervenors to raise issues repeatedly and thereby prevent closure. These "dilatory tactics" cost the utilities money and time without improving anything. Indeed, the objective was not "any genuine safety or environmental concern, but rather stalling the process long enough to force" a project's "abandonment." "In the case of the Seabrook plant," Copulos claimed, "this has been especially evident."

However certain his prose, Copulos's attribution of motive had no obvious evidential base and failed to differentiate among several groups of nuclear opponents. Probably his discussion of delay as a tactic was supposed to describe such organizations as SAPL and the New England Coalition on Nuclear Pollution (NECNP), both of which followed legal procedure and used rules to their purpose when possible. In fact, however, neither group invariably sought delay; both did stall when it seemed advantageous to do so, as did PSNH on occasion.

But when Copulos discussed the Clamshell Alliance, he made his target specific and left little to inference. Clams, he wrote, "are not merely opposed to nuclear energy; they are opposed to the very way in which most people live their lives." Thus this was no mere struggle over a generating station; what was at issue was nothing less than the American way of life. Clams "would impose . . . a steady-state economy in which no economic growth" would occur. Their "emphasis on conservation and the utilization of labor intensive methods of manufacture would," Copulos predicted, "have major impacts on our current mode of living if implemented." The alliance was not only economically regressive but socially authoritarian as well, for without consent it would fasten "its value system on society as a whole." The Clams' attitude betrayed a "rather elitist strain" that was "contrary to the fundamental underpinnings of our democratic system" and would bar "upward mobility . . . thereby depriving women, the young, and members of minority groups" of opportunity. Copulos had reversed the image Clams had of themselves. They took pride in participatory democracy; Copulos found authoritarian bigotry. They looked ahead to a decentralized economy that provided full, nonalienating employment; Copulos thought their vision clouded and reactionary. His argument had the virtue of consistency, for he began and concluded with the postulate that the Clamshell Alliance had a skewed view of American society. But his demonstration, for all the precision of his statistics, rested on his own claim, and that of his sponsors, to define norms for other Americans, and he thereby displayed a social arrogance like that he deplored in the Clamshell Alliance.[30]

While the Clams flared in demonstrations and expired in division and futility, SAPL doggedly stuck to its legal strategy. When resources allowed, and often when they did not, SAPL resolutely challenged the utility's plans and appealed regulatory decisions. The organization did not achieve great public visibility, even after the media's preoccupation with the more colorful Clams subsided. The Clamshell Alliance had made nuclear opposition temporarily untenable; Backus once remarked to Chichester that the Clams' excesses had paradoxically served PSNH well, strengthened the resolve of the plant's proponents, and probably contributed to the decision to press forward with construction.[31] Though the Clams' preeminence was not the only reason, SAPL was not able to take advantage of several events in 1978 and 1979 that ought to have been inspiring.

In 1978 the New Hampshire legislature prohibited the Public Utilities Commission from adding the cost of Seabrook's construction to the base used to calculate consumer rates. To add Seabrook's cost before the station produced electricity, the legislature decided, would make PSNH's customers involuntary investors in the project, which ought to be financed by stockholders, earnings, and debt. Governor Thomson, predictably, vetoed the bill, but the issue gave the campaign of his Democratic opponent, Hugh Gallen, an immense boost. After Gallen's election in November, the legislature revived the bill, which the new governor signed in May 1979, two months after the accident at Three Mile Island paralyzed the nuclear industry.

At the time, PSNH appeared to shrug off these twin disasters, though the failure to include Seabrook in the rate base and the regulations on emergency planning the NRC adopted as a result of Three Mile Island (TMI) would eventually take on enormous significance. More surprisingly, the events did not revitalize SAPL, which held a rally and a press conference, neither of which attracted much attention, and conducted fund raising out of habit and without much result. Immediately after TMI, Backus had sought revocation of the construction permit because of the absence of emergency plans, a motion that was denied but would have legal echoes. By 1980, although Backus had suggested several new initiatives, the league had no money to pursue them and little sense of organizational urgency. With justifiable petulance, the executive director protested the board's inaction, noting that no one had called or planned an annual meeting. In the winter of 1980–1, the directors voted to reemphasize the league's original effort to defend the seacoast environment from acid rain, toxic waste, and inadequate zoning, as well as nuclear power, a stance intended to diminish organizational concentration on Seabrook.[32]

However compelling those other causes, they had less appeal to SAPL's constituency than the continuing legal battle against Seabrook Station. By 1981, given a financial respite with a grant from the family foundation of President Anne Merck-Abeles and energy and organizational stability by Jane Doughty, the new executive director, SAPL returned to a policy of consistent legal obstruction of every effort to operate the plant. Fainthearted opponents, Merck-Abeles wrote, thought the struggle had been lost several times, but SAPL was going to stay the course; the fact that the plant was about three-quarters finished, she asserted, "doesn't make it any less wrong." Backus gave this determination legal expression in challenges to the company's financial viability, to its managerial reorganization, to the design of the plant's control room, to the use of chlorine in the tunnels, to New Hampshire's emergency plan, and to anything else that might facilitate operation. As construction concluded and operation still seemed months, and perhaps years, in the future, SAPL remained vital to opposition that became almost fashionable as it expanded. The league continued to owe Backus money, and Jane Doughty's salary was usually in arrears, but both of them, and the organization they came to symbolize, had, with talent, persistence, craft fairs, and twenty-dollar donations, prevented operation until a large fraction of the public, and some important politicians, reached SAPL's state of antinuclear environmental consciousness.

Some of those politicians, especially a few from the New Hampshire seacoast, had enlisted early. Yet a handful of representatives could not overcome the inertia of the state's huge legislature without assistance and guidance from the executive branch. One consequence of New Hampshire's refusal to tax is political leadership that often lacks time and sometimes expertise to master new, complex, and contentious topics like nuclear power. And, until emergency planning surfaced in 1981, local leaders had no opportunity effectively to participate in decisions about Seabrook Station, which was a matter for federal and state, not local, government. Even when emergency planning moved to the center of the controversy in 1984, many local leaders chose to ignore the problem rather than expose their towns to legal expense and themselves to hours of uncompensated effort.

Nevertheless, the cohort of anti-Seabrook legislators grew, and several towns did contest the state's emergency plan. More important, the Commonwealth of Massachusetts, which had participated in the licensing process from the outset, became a major player in 1986 when Governor Dukakis declined to submit any emergency plan at all. Before

Seabrook Station could operate, then, owners were likely to have to prevail in a court battle, not with some underfunded intervenor, or even with a frugal New Hampshire town, but with Massachusetts. Citizens' groups, including SAPL, NECNP, and the Clamshell Alliance, had created enough time for the political climate to change and opponents more formidable than the citizens' groups themselves to emerge. The accident at Three Mile Island and the catastrophe at Chernobyl gave credibility to opponents who had once seemed eccentrics operating on society's fringe. If the discussion of safety had remained focused on environmental or engineering details and the niceties of design, the interest of the public and politicians might well have been less intense. But emergency planning, which involved schools, dangerous intersections, elderly parents, and traffic at the beach—these were matters everyone and their elected representatives could grasp.

Thus emergency planning broke the monopoly of expertise and made nuclear power a broadly intelligible matter. When sensible people concluded that plans were defective, official reassurance rang hollow and, when repeated, discredited those who reiterated it. Emergency planning moved alleged ineptitude out of the plant, where it was unseen and unverified, to the neighborhood, where it was entirely evident. As the final item on the regulatory agenda, emergency plans were aired before a much larger audience, with more personal concerns, than those that predominated when all that was at stake was somebody else's environment.

PSNH and its economic and political allies—governors, labor leaders, editors, investment bankers—had routinely blamed regulatory delay on intervenors. In fact, the company itself had asked that consideration of emergency plans be postponed, a decision that must have seemed in retrospect a tactical blunder. In this instance, intervenors certainly secured unexpected advantage from a delay they welcomed but did not cause. Nor did they cause the other interruptions at Seabrook with which they were charged. Those missed deadlines achieved so much notoriety in part because the nuclear establishment advertised them and in part because they were so numerous. The regulatory process, of which intervenors took advantage, surely lengthened a construction schedule that slipped beyond recognition, but so did weather, design deficiencies, and financial stringency. SAPL and the Clamshell Alliance and the Commonwealth of Massachusetts wrote neither the unrealistic schedule nor the NRC's regulations and in fact objected to more of the latter than did PSNH. The utility, however, and its allies needed the NRC's approval, and that imperative no doubt muted their criticism of the agency and its rules. For want of the real target, the intervenors had to serve. They took the blame proudly.

After the catastrophe at Chernobyl, Stephen Comley decided to investigate arrangements to protect residents of the nursing home he owned in Rowley, Massachusetts, from a nuclear accident at nearby Seabrook. There was, he found, no plan for the Sea View Nursing Home, not only because Massachusetts had declined to submit emergency plans but also because his property was two miles beyond the ten-mile distance for which emergency planning was required. Further, he discovered that emergency procedures for other communities in the area did not insure transportation and support for those unable to care for themselves. That news, one of his friends remarked, launched Stephen Comley "like an unguided missile."

Comley fired off letters to newspapers, deploring public policy that appeared to protect those able to "run when the siren goes off" and to ignore those who most needed assistance. He urged Governor Dukakis to require plans for Rowley because children from the town attended school in Newbury, where they would be required to follow evacuation procedures. He wrote petitions that secured thousands of signatures. He went to Washington to press his argument on the Nuclear Regulatory Commission, the president, and members of Congress. He hired airplanes to carry his message over government buildings in Washington, Boston, and Concord, New Hampshire.

He was an unlikely lobbyist, this stocky, forty-two-year-old Republican from Rowley. He had voted for Ronald Reagan and contributed to the Republican senatorial campaign chest. But he was too impatient, too pugnacious, to bother to be ingratiating, and he tended to open conversations with "Why the hell . . . ?" instead of some more tactful inquiry. "What the hell kind of a way is that to treat your . . . grandmother?" he asked *Washington Post* columnist Mary McGrory after describing the emergency plans' provisions for the infirm. But, McGrory noted, "none of the smart people" Comley pestered in Washington would answer him; with the exception of NRC Commissioner James Asselstine, official Washington brushed him off.

Comley returned to Rowley, filed more petitions, advertised his search for whistle-blowing former construction workers and his willingness to provide them legal support, and urged the voters of New Hampshire to retire Governor John Sununu, whose support for Seabrook Station rivaled Thomson's, in the election of November 1986. When too few voters paid attention, he tried to deliver his antinuclear lecture in person at Sununu's inauguration. He was arrested, charged with disorderly conduct, and convicted. He appealed and vowed to continue his effort to prevent the operation of Seabrook Station until he saw adequate protection for people for whom he felt responsible.

Stephen Comley was a one-man band. He kept his own counsel, devised his own campaigns, and spent his own money—more than $100,000 by his own accounting. He played no role in any of the dozens of organizations that shared his objectives. By his own admission, he came late to a realization of nuclear hazards, and he tried to atone for tardiness with frenzy. Once persuaded, he believed his case so self-evident that it needed only articulation to carry the day; if the governor or the NRC or the president only knew what he knew, Comley believed, the plant would not be licensed. When he was not allowed to deliver his message in the usual ways, Comley employed less conventional means, not all of which seemed characteristic of a sober Republican business-man. [33]

Robert Preston, a sober Democratic businessman, registered his opposition to Seabrook Station over a decade in public life. The minority leader of the New Hampshire Senate, Preston's stance was unambiguous before his first campaign and before the views of his Hampton constituency had hardened. His oratory was neither flamboyant nor inspiring; he was, he said, no table pounder. He did not run against PSNH and Seabrook Station; indeed, he once remarked accurately and diffidently in the middle of a low-key campaign, he could be reelected without discussing the topic. But he opposed the utility's scheme from the moment it involved a canal through the marsh, and he grew increasingly disenchanted with the way various levels of government had dealt with the serious environmental and safety issues the plant posed. He had witnessed the sly effectiveness of PSNH's legislative lobbyists and listened to speeches they prepared for some of his fellow legislators. He worried because the state seemed to amplify a corporate message more loudly than the voice of the people. He observed that the federal government's regulators seemed hard-of-hearing and their rules gave the appearance that "the deck is stacked." Although he had impeccable establishment credentials, including graying hair, three-piece suits, and a plaque noting his presidency of the Chamber of Commerce, Preston said he guessed he was part of the "small vocal minority" some of his senatorial colleagues derided. Preston disparaged neither the noise that some young opponents of the plant generated nor the idealism that he believed motivated them.[34]

Some once-young opponents aged as the struggle dragged on. Rennie Cushing, the Clamshell activist who was arrested several times in his twenties, continued his protest in his thirties in the New Hampshire legislature as a representative from his hometown of Seabrook. Robin Reed, a friend of Cushing's from their days in the alliance, joined him in the legislature representing Portsmouth. Sharon Tracy, who as a college undergraduate had written a good deal of the copy for the *Clamshell Alliance News*, later wrote press releases for a three-state coalition of state officials, including Senator Preston, who tried to block operation of the plant. Herbert Moyer joined SAPL when he was a young high school teacher fresh out of the University of New Hampshire. After a dozen years of quiet, effective organizational service in several opposition groups, he was arrested at the Exeter dump while collecting signatures opposing the attempt to reduce the ten-mile radius for emergency planning. He won the resulting court case, monetary damages, and election as an Exeter selectman. He gave the financial settlement to the American Civil Liberties Union. Guy Chichester, still trying to raise the region's political consciousness a decade after the collapse of the Clamshell Alliance, devised media spots to combat those of public relations fronts sponsored by PSNH.

Grafton Burke, Jr., served his antinuclear apprenticeship unwittingly as a construction worker at the plant that rose within a mile of his rented quarters in Seabrook. He observed what he took to be irregularities in construction, to which he pointed in conversations with foremen, quality control people, and NRC inspectors. His persistence as a self-appointed monitor did not endear him to those with whom he worked or to those to whom he reported. The resulting hostility, he believed, found expression in intimidation and eventually led to his being fired.

Burke came to see himself, his family, and others associated with him as victims of PSNH. The company, he believed, manipulated labor unions, public agencies, physicians, courts, and individuals in a concerted campaign that included arson, perjury, assault, and involuntary detention as tactics to overcome opposition. He thought an accident, similar to Karen Silkwood's fatal crash, had silenced the potentially damaging testimony of a welding inspector, who in fact was alive and able to plead guilty to charges that he falsified his reports. Burke spent some weeks in a psychiatric hospital, and his testimony betrays occasional confusion and a view of the world that might be called paranoid.[35]

That perspective was not unique to Grafton Burke, and not necessarily irrational, though most Seabrook foes did not invest PSNH with the omnicompetence Burke attributed to the corporation. He believed he was engaged in a David-and-Goliath struggle and that PSNH and its allies would use any means to crush him and other ordinary citizens who stood in the way. However unusual his angle of vision, the hostility and mistrust Burke felt for PSNH were not unusual at all.

Late in September 1986, for instance, Judge Sheldon Wolfe tried to pacify unruly spectators at a hearing called to consider on-site emergency planning. Although Wolfe had warned correspondents that the issues were technical and that the rules forbade public participation, he encountered a poisonous reception that was clearly not conducive to an adjudicatory proceeding. Voices demanded that the microphones be adjusted, informed the audience that police were towing vehicles in the parking lot, and interrupted Wolfe's attempt to outline the agenda. Placards waved silent protest, which did not long go unspoken. When Wolfe said the board would take no public statements, someone taunted, "therefore we can sit here and shut up and just watch you execute us."

Forced to the realization that he would have to permit participation or have no hearing at all, Wolfe gave in. He allowed what the NRC called limited appearances, unsworn statements of opinion that were not weighed as evidence. The opinions were firmly held, forcefully stated, and invariably hostile to PSNH, the NRC, and nuclear power. Calling herself Karen Silkwood, a woman from Keene attributed her five miscarriages to a former residence near a nuclear test site, and she vowed to be arrested several more times if necessary to prevent nuclear power from becoming established in New Hampshire. A physician protested the "steamroller" of federal power that had subverted "local control" and disposed people to give in to "totalitarian forms." Other witnesses sang songs, read from the Book of Revelation, or masqueraded as Indian sages and railed against the white man's dangerous technology.

It went on all day and through two additional evenings, and it was not entirely hyperbole and hysteria. But it was all angry, even when the rage was under control. Barbara James, who had done yeoman duty in several antinuclear groups, avowed her growing irritation. "Many of us," she said, "have been at this for more years than any of us want to think about." And she was "very tired" of being told that "we should come up with a workable evacuation plan," when PSNH had been "paid well to create the danger."

> I am angry at the audacity of them to expect us to take all the risk and do their work for them as well. . . . I would like to send a bill to the Public Service Company at their hourly rates for all the hours over the past twelve or more years that we have all spent on Seabrook.

The energy that sustains a commitment like that, Dr. Robert Du Pont has said of the antinuclear movement, stems from fear. A professor of medicine and public health at Georgetown University, Dr. Du Pont has contrasted the intensity of the public response to nuclear power with that to cigarette smoking. On the one hand, he points out, no one has died in the United States as a result of the commercial production of electricity with nuclear fuel; on the other, several hundred thousand people have perished because of tobacco use. Yet the public response is inversely proportional to the danger, a fact Dr. Du Pont explains by citing four factors that escalate the fear of nuclear technology: fear of the un-familiar; fear of the absence of personal control; fear of the instant catastrophe (as opposed to the creeping disaster); and fear of a force for which there is no perceived need. Those factors heightened apprehen-sion, he explained, and although people who displayed such symptoms were not irrational, deluded, or ill, they did exaggerate the threat beyond reason.

The diagnosis was no doubt accurate, although Du Pont's prescrip-tion—an educational campaign to highlight the industry's record of safety—seemed a bit like treating cancer with public relations. And, however carefully qualified and however accurate, Du Pont's explana-tion seemed to antinuclear activists one more example of expert opinion that varied from their own experience. For fear was only one cause, and not always the most important one, of their response to the nuclear juggernaut. Or at least fear was not simply concern about radiation or a sudden explosion. Opponents of nuclear power also feared damage to the balance of nature and to self-government; they feared economic, political, and centralized power, as well as nuclear power; they feared corrupt governmental process, technical incompetence, and human arrogance.

And the energy of the movement did not result entirely from fear of anything. "I've come to realize," Janet Schaeffer remarked during Judge Wolfe's Portsmouth hearing, "that people in Washington must think that what motivates people here to be against nuclear power is fear."

Those same people, she supposed, figured that "our opposition will evaporate" if "you can somehow alleviate those fears." But that estimate, she said, did not describe the people she knew, who were determined rather than afraid, who were informed and not ignorant, who expected to prevail and did not fear defeat, and who were, above all, full of hope. [36]

4 | Money and Management

"Teeter[ing] precariously on the brink of financial disaster."

The nuclear industry was almost two decades old when the Public Service Company of New Hampshire first proposed to build a generating station at Seabrook. Although early cost estimates in the industry had been disastrously unreliable, utility managers, in New Hampshire and elsewhere, maintained that each new facility would become the first exception to the rule. Even in the 1960s, before energy costs, interest rates, and inflation made most economic forecasts suspect, a trade association spokesman admitted that "estimating capital costs for power plants is like shooting at a moving target." The *Technology Review* despaired of hitting it: "We just have little firm idea of what the actual cost in deflated dollars of reactors ordered subsequent to 1969 will turn out to be." Perhaps larger reactors, which would spread capital costs across increased output and thereby reduce investment per kilowatt, or a learning curve that would hold labor costs down, or some technical improvement or regulatory reform would bring actual expenditures for construction into line with estimates. Meanwhile, until costs stabilized, some suppliers appeared deliberately to have submitted unrealistically low bids for nuclear projects based less on predicted costs than on comparisons with competitive fuels. That is, nuclear contractors and suppliers may have used pricing as a means to the sale, knowing that ultimate costs would exceed figures on which utilities based their first budgets.[1]

These practices were not trade secrets, because the cost overruns of regulated monopolies could not be concealed. Yet utilities in the 1960s and early 1970s flooded manufacturers with orders, promising skeptics that the partnership between technology and free enterprise would soon produce cheap, safe nuclear plants on American assembly lines. There was no experiential base for that reassurance; neither the learning curve nor enhanced efficiency nor standard designs had occurred, and costs were clearly rising instead of falling off. But a bandwagon psychology gripped managers of the nation's public utilities, who decided they could not afford to miss the promising nuclear future. The vice-president of one company observed privately that "the average utility knew as much about the nuclear plant it was buying as the average car-buyer knows about cars." Managers knew how big the plant was and what somebody said it might cost. "We got into nuclear power," he continued, "because the president of my utility used to play golf with the president of another utility. They bought one, so we bought one."[2]

PSNH made one false start. But given a second chance three years later, New England utilities oversubscribed a project more than twice as large. The new project, PSNH predicted, would cost less than a billion dollars and start producing in 1979. The company was convinced that "delays and other problems experienced by other nuclear plants won't happen at Seabrook." By 1979 the industry would have removed impediments to efficient construction and prompt operation: "we expect the bugs to be worked out by then," PSNH vice-president Elliot Priest told a New Hampshire reporter. "The way they're working on them, the solutions will come any time now. That's the nature of this business. A problem comes up and we solve it."[3]

PSNH lacked the financial foundation for that sort of confidence. In 1972, when Priest was predicting quick solutions to all construction problems, the company had less than $400 million in total assets and more than $200 million in debt. Net income approximated $11 million, of which dividends on common stock required more than $9 million. Of the roughly $2.25 million remaining, more than $2 million consisted of noncash income called allowance for funds used during construction (AFUDC). The company had a net annual positive cash flow of about $100,000, therefore, and proposed to undertake half the cost of constructing two nuclear reactors.[4] By the following summer, when the Atomic Energy Commission (AEC) received the initial application, the estimated cost of PSNH's share of Seabrook Station had crept up to $600 million.

Clearly, the company's earnings could provide only a small fraction of the required capital. Although the New Hampshire Public Utilities Commission (PUC) would no doubt permit an increase in rates, and completion of other construction would free corporate funds for Seabrook, most of the money to build the plant would have to come from outside sources. A bond prospectus late in 1974 noted that the company would need almost $600 million to meet planned capital expenditures between 1974 and 1978. "Cash available from internal sources . . . is expected to aggregate about $77,000,000."[5] The remainder—a sum larger than the company's total assets—would have to be raised in the capital market through the sale of stocks and bonds and through short-term loans.

Nineteen seventy-four was not a good year for utilities. Electric rates lagged behind both general inflation and the escalation in oil prices precipitated by the Arab boycott, a change that was directly relevant to PSNH, which relied on oil for most of its electrical output. Rate-setting agencies, which had become accustomed to reducing charges in the prior decade or two, reacted slowly to the changed economic environment. In April 1974 Consolidated Edison, a bellwether holding company in New York, announced that it would not pay a quarterly dividend for the first time in eighty-nine years. The announcement, one observer wrote, "hit the industry with the impact of a wrecking ball." By September, the average price of utility shares had dropped more than a third, and the interest rate on bonds had risen sharply.[6]

Uncontrolled inflation, rising interest rates, and shattered investor confidence were not propitious circumstances for a small utility trying to borrow a great deal of money. A share of PSNH's common stock, which had a book value of more than twenty dollars and which had sold as recently as 1969 for more than thirty dollars, netted the company less than twelve dollars when a million shares were sold in October 1974; bonds issued at the same time carried an interest rate of 12.75 percent.[7] But management used quite different numbers as the basis for internal financial plans. Just two months later, PSNH assumed that stock could be sold at book value, when in fact it had recently brought only 60 percent of that amount, and that bonds would bear interest at 8 percent, when experience suggested that the actual cost would be almost 60 percent higher. Further, the company assumed that the Seabrook construction schedule, which had already begun to slip, would be kept and that inflation, which was approaching double digits, would never be greater than 8 percent. Most of those assumptions were demonstrably contrary to fact when they were made, and financial projections predicated on them were at best curious fiction and at worst absurd.[8]

Inaccuracy was only one peril. The company's financial data had to support two contradictory interpretations. On the one hand, the outlook had to be sufficiently rosy to persuade federal regulators that PSNH was able successfully to undertake half of the Seabrook project. Similarly, potential investors had to believe that both capital and return were secure. On the other hand, the New Hampshire PUC had to see enough financial hardship to permit rising rates and profits, so that both the construction schedule and the debt structure could be maintained. This financial high-wire act, performed for the most part by Robert Harrison, PSNH's financial vice-president and President William Tallman's eventual successor, went on for most of a decade. Finally, in 1982, the Nuclear Regulatory Commission (NRC) ruled that PSNH was financially qualified and the issue was not subject to further litigation; ironically, by that time the company was in fact essentially insolvent, a condition that became increasingly evident even if it no longer seemed relevant to federal officials.

A formula for reconciling these contradictory interpretations of financial data evolved early. The company conceded that its ambitious construction program depended upon favorable rulings by the PUC. But, PSNH argued, such rulings could reasonably be anticipated, because the commission had endorsed Seabrook Station; acceptance of the project in principle implied a willingness to grant the rising rates and profits necessary to complete it. "Were this not so," PSNH asserted, "power supply to the homeowners and industries in the state would be in serious jeopardy." The company was at least outwardly confident that the PUC would recognize almost a shared responsibility to encourage investment in the nuclear plant that would meet the state's electrical needs. That confidence did not seem misplaced when, at the end of 1974, the PUC granted a rate increase calculated to produce more than $17 million annually and a 14 percent return on investors' equity.[9]

That decision persuaded the NRC. Tim Jackson, a consultant engaged by the commission's staff, relied heavily on the ruling as he prepared to testify before the Atomic Safety and Licensing Board (ASLB) that would rule on the construction permit. Jackson expected new rates and higher profits to raise the price of PSNH stock to approximately book value and to reduce the rate of interest on bonds to about 8 percent. Recognition by investors of the company's improved prospects would make Harrison's projections reasonable. In Jackson's view, then, PSNH was "financially qualified," as the NRC's regulations required.[10]

Those regulations, it developed, were not very specific. A staff member subsequently testified that there were "no guidelines written down" for assessing an applicant's financial qualification. Indeed, because "the financial picture changes over . . . time, and the requirements of a particular company change," the witness continued, "the financial area does not lend itself to ironclad rules." In other words, qualification was a matter of judgment, and the staff and its consultants in 1975 believed PSNH had the financial strength to proceed. No one apparently wondered whether a smaller slice of the project might have been more manageable than the 50 percent share PSNH had undertaken. [11]

Robert Harrison conceded that raising nearly twice the company's 1974 assets in eight years might be difficult, but he had no doubt about the outcome. PSNH had, he noted, nearly accomplished that feat in the preceding eight years, although market conditions, to be sure, had been more propitious. Still, he stuck to his prediction that he could sell stock at book value and debt costing the company 8 percent. The most recent bond issue at 12.75 percent, Harrison said, had been a temporary aberration, which the company's financial health would soon correct; over time, the interest rate on PSNH bonds would average about 8 percent. How much of the company's outstanding debt bore interest higher than that? asked an attorney for one of the intervenors. Harrison thought the answer was about half, an estimate that implied a remarkable decline in the immediate future to achieve the 8 percent average he predicted.

Attorneys for opponents of the plant also inquired about Harrison's plans to sell PSNH stock. Had the ratio of market value to book value not declined without interruption from 170 percent in 1968 to about 50 percent in 1974? Yes, Harrison said, but a few years did not constitute a trend, and the PUC's recent ruling had already boosted the stock's market value. Selling a new issue at 50 percent of book value, as the company had done in 1974, was, Harrison admitted, "catastrophic," but the catastrophe had passed; PSNH stock had recently advanced while shares in other utilities had declined. As long as PSNH's earnings and dividends increased, Harrison told the chairman of the licensing board, the market price ought not to fall below the value per share shown on the company's books.

Harrison's confidence was not infectious. Was it customary, asked Ernest Salo, a member of the ASLB, for utilities to borrow twice the total of existing assets? Not usual, Harrison replied, but not unprecedented. Was the company now earning the 14 percent return the PUC allowed? Not yet, Harrison said, but he expected to see that result soon. How come the company had decided to build Seabrook when a smaller project had been discarded only a couple of years before as too expensive? Circumstances change, Harrison responded. Did his cost figures include the most recent modifications of the cooling system? He did not know much about the technical specifications of the plant, Harrison answered, and he did not adjust his numbers for every design change. What sort of calculations had he made to reach the assumptions on which his financial arrangements depended? "I did not," Harrison said, "perform any specific calculation." Instead, he had relied on "what I call my experience and knowledge in the field."[12]

Other experts with different experience (or different employers) had different ideas. Professor James Nelson, an Amherst College economist who had studied regulated industries extensively, testified for the New England Coalition on Nuclear Pollution (NECNP). Nelson compared the data submitted by PSNH to those of its partners in the enterprise, several of which seemed to have much stronger balance sheets. Nelson pointed to "marked discrepancies" among the partners "as to the terms on which they can raise capital." Further, he discerned "a paradoxical, if not perverse, tendency for companies which now have the lowest credit ratings to assume that they can raise capital on the easiest terms." He thought PSNH's recent experience was enough to belie Harrison's cheerful financial forecast, because Seabrook Station, which would represent most of the company's assets, would produce neither electricity nor profit for years; the investment, in other words, would inevitably be sterile for some time, a condition that seemed to Nelson unlikely to attract investors. Commitment of so much of the company's resources to one project, Nelson held, "amounts to putting a high percentage of one's eggs in one basket." And selecting nuclear fuel for the plant, with all the attendant uncertainties of cost, regulation, and reliability, compounded the problem by "deliberately selecting a chancy basket."

Thomas Dignan moved to eliminate Nelson's remarks about nuclear power as beyond his expertise. Dignan also demonstrated that Professor Nelson had made several overstatements and a few miscalculations. But cross-examination affected the edges of Nelson's remarks and did not shake the central point: Harrison's assumptions were unduly optimistic, and the undertaking was very risky for a small company like PSNH. Purchasers of the company's securities, Nelson said, would, in the case of common stock, be buying paper representing "no cash yield at all" but rather earnings "due essentially to peculiarities of public utilities accounting."[13]

Nelson referred to the allowance for funds used during construction and perhaps to depreciation, both of which could add substantially to a utility's reported earnings but neither of which represented cash in hand.

AFUDC is a complex accountant's device to allow a utility a return on capital invested in a plant that is not yet productive. In effect, AFUDC adds both to net income and to the cost of construction. "Instead of charging off interest costs" as an ordinary business might, explained James Cook in *Forbes* magazine, "the utility imputes an interest charge to the money it has invested in the project, capitalizes the amount (that is, adds it to total investment), and reports it as a non-cash credit in its income statement." The result, in the case of a $2 billion investment, would be an increased cost of $200–300 million per year, which then compounds. Thus, Cook continued, the longer a plant is unfinished, the more the cost compounds, the larger the eventual addition to the rate base, and the higher the percentage of earnings becomes that is not real cash but only an accounting entry. Harrison's projections did not have to run many years before AFUDC constituted more than half of PSNH's earnings. [14]

Marvin Mann, a member of the Seabrook licensing board, found this convention very puzzling. Would not people with capital discern, he asked Robert Harrison, that AFUDC was simply a way "to inflate advertised earnings"? Indeed, what other "practical effect" could the device have, since it did "not generate cash money"? Harrison said the practice was not deceptive, but on several other occasions he confirmed Mann's guess and Nelson's assertion that sophisticated investors simply deducted AFUDC from statements of earnings. By 1985 AFUDC amounted to almost half of PSNH's reported $2 billion investment in Seabrook Station and more than 100 percent of the company's earnings. [15] That predicament had not been impossible to foresee a decade earlier, as Nelson's testimony demonstrated.

David Lessels, the finance director for the New Hampshire PUC, knew that financial exigencies would prod PSNH to seek continuous rate adjustments. He quickly gathered statistics for the commission's guidance. By no means definitive, Lessels's study nevertheless showed that rates had doubled in four years, that PSNH had consistently over-estimated demand for electricity and therefore its own revenue, and that the generating capacity it proposed for Seabrook was not needed and would boost without cause the electricity bills of New Hampshire consumers. Finally, he believed that the utility's estimate of costs for Seabrook Station erred by about a billion dollars and that the investment was precarious as well as unnecessary.

Those conclusions were unwelcome at the PUC, which pigeonholed Lessels's report, and anathema in the governor's office, where Meldrim

Seabrook Station. Photo courtesy of New Hampshire Yankee

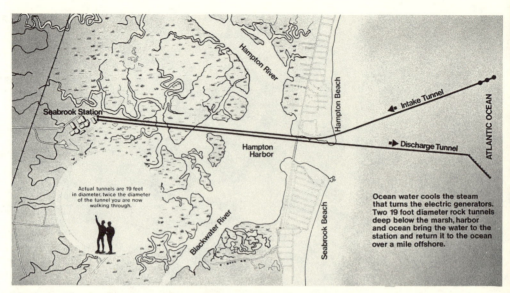

Actual tunnels are 19 feet in diameter, twice the diameter of the tunnel you are now walking through.

Ocean water cools the steam that turns the electric generators. Two 19 foot diameter rock tunnels deep below the marsh, harbor and ocean bring the water to the station and return it to the ocean over a mile offshore.

Marsh, beaches, and cooling system. Graphic courtesy of New Hampshire Yankee

Outer shell of 1150 mega-watt reactor. Photo courtesy of New Hampshire Yankee

Containment building for Reactor I. Photo courtesy of New Hampshire Yankee

Audience at ASLB hearing, 1986. Photo courtesy of Greg Bergman, Rockingham County Newspapers

Opposite page: Robert Backus makes a point;
Judge Helen Hoyt at left.
Photo courtesy of Herbert Moyer

PSNH attorney Thomas Dignan.
Photo courtesy of Herbert Moyer

Boston Sunday Globe, July 10, 1988.
Reprinted with the permission of George Hammond

Copyright © 1988, Boston Globe. Wasserman cartoon,
reprinted with permission of Los Angeles Times Syndicate

Photo courtesy of New Hampshire Yankee

Demonstration, 1977. Photo courtesy of Steve Bromberg

Photo courtesy of Herbert Moyer

Thomson was about to order state employees who did not support the project to remain silent. (A few months earlier, in urging the NRC to issue a construction permit promptly, Thomson had pointed to the financial hardship caused by delay. "Recent estimates," he wrote, "indicate that delays . . . are adding at least $10 million each month" to the cost of Seabrook Station. The estimates were fantasy; construction had not begun, and PSNH's entire investment in the project might have reached the $10 million Thomson said was accruing each month.) Robert Backus, who represented two intervenors in the federal licensing process, sought to compel Lessels to disregard the governor's "gag order" and to share his work with the ASLB. But Dignan persuaded the board not to issue a subpoena on the grounds that Lessels was no expert on economics. The appeal panel subsequently observed that the ruling had been clearly wrong but was essentially harmless, since other evidence demonstrated that the power was needed and that PSNH could meet its financial responsibilities.[16]

That evidence included "the preponderance of expert testimony," as the ASLB weighed it, "that the necessary funds will be forthcoming" even if Robert Harrison had not accurately predicted interest rates and stock prices. The board's assurance stemmed from state control of electric rates, "a vital factor affecting the ability of applicants to finance Seabrook Station." In effect, the judges abdicated to a state agency the task of assuring financial qualification that NRC rules assigned to them. The board did not refute Nelson's testimony so much as it ignored him in ruling that construction at Seabrook might begin.[17]

But PSNH was in no position to finesse the problem because, as Lessels had predicted, the company had to keep returning to the PUC for more revenue. Every such occasion created a record documenting out of the mouths of officers and consultants the utility's financial distress; intervenors copied the statements and mailed them off to the NRC with motions raising once more questions about PSNH's financial stability. In particular, Robert Harrison sought to convince New Hampshire authorities to allow PSNH to include in the rate base funds expended to construct Seabrook Station—or construction work in progress (CWIP), in the jargon of utility accountants. The device, which New Hampshire had never permitted, allowed the company to receive a return on investment in new generating capacity before it became productive, thereby forcing consumers to provide part of the requisite capital. In 1973 Harrison had said PSNH would not need this regulatory concession to complete Seabrook Station; in 1977, twelve months after ground was broken, he forecast brownouts and economic disaster for the state as well as the company unless CWIP was immediately added to revenue.

Harrison's investment bankers corroborated his judgment. William Harty, of the Morgan Guaranty Bank, reviewed several financial schemes in connection with borrowing PSNH proposed. None, he said, made economic sense without the additional income CWIP would provide. Eugene Meyer, a utility specialist with Kidder Peabody and Co., asserted that PSNH could not raise sufficient capital from operations to attract outside investment. The dividend in 1976, Meyer pointed out, exceeded the company's net income if AFUDC were excluded. James Nelson had previously suggested that sophisticated investors would make that subtraction, and, although Meyer's emphasis differed, his testimony confirmed Nelson's work in other respects as well.[18]

The New Hampshire Public Utilities Commission sought an independent opinion about PSNH's need for revenue based on incomplete construction. The commissioners engaged Daniel Trawicki, whose investigation showed once more that PSNH had undertaken a very large project on a very small financial base. Although the PUC had permitted a 14 percent return on equity, Trawicki noted, the company had not been able to earn that much and consequently sought an increase of nearly 30 percent in the rate base for facilities under construction, as well as a further increase in rates. Trawicki estimated that CWIP, if allowed, would provide almost half of the company's earnings in 1978 and about 70 percent three years later; without CWIP, there would be no cash earnings, just as other observers had predicted. The company's construction program would require annually about 70 percent of total assets—an enormous requirement that Trawicki believed could only be met through CWIP: "I do not believe that it is in the best interest of ratepayers or investors to have the company teeter precariously on the brink of financial disaster, not through any fault of its own but in order to meet the obligation to provide safe and adequate service." The PUC concurred and authorized CWIP in the spring of 1978.[19]

While PSNH teetered in Concord, it had to maintain its poise at the NRC. One balancing tactic was to obscure the contradiction between the two postures with tactical objections designed to keep what was said in Concord out of the NRC's record. When he had to deal with substance, Thomas Dignan, often joined by counsel for the NRC staff, argued that financial plans PSNH submitted to the federal agency were illustrative and nonbinding and that the company had to be able to alter details at discretion. Further, Dignan maintained, the NRC had no cause to examine the financial health of regulated public utilities, whose revenues were subject to the oversight of state agencies. If a company was regulated, Dignan asserted, then it was by definition financially qualified; federal officials had simply to "check the box." Although Dignan claimed that was "as true as God made little green apples," his check-the-box test was too automatic for adoption by the NRC.

The Seabrook appeal board asked Dignan if an applicant could ever be financially unqualified. "I don't think we can or should," Dignan replied. Then, Michael Farrar of the appeal board continued, perhaps financial inquiries were altogether unnecessary. "No," Dignan answered, "It's not a non-issue."

> You could have a situation [although] I don't think you have it here, . . . where, say, some utility came in, bit off a big chunk of nuclear, and you took a look at it and, you know, the thing was either bankrupt or close to it, I think you'd have to deny them a license.

The point of the requirement, Dignan supposed, was to prevent a company from cutting corners during construction or starting a reactor without adequate security or running out of money while a turbine was still spinning. That was not the condition of his client, which was meeting its payroll and paying its bills, and which would secure further rate increases as future needs developed.

Attorneys for intervenors were unconvinced. If any utility anywhere was ever to be found unqualified, as Robert Backus put it, this seemed to be the occasion. The attorney general of Massachusetts agreed that a drooping bond rating, an evaporating line of credit, and a failure to earn the allowed rate of return seemed curious proof of financial qualification. The company's position, argued the New England Coalition on Nuclear Pollution, "totally defies logic." PSNH had submitted Trawicki's report to demonstrate that New Hampshire regulators would permit CWIP and thus buttress the company's financial position. But that investigation, of course, also revealed PSNH's "severe financial distress," which the company paradoxically seemed to claim constituted financial qualification. "In truth," counsel for the NECNP observed, "we are at the Mad Hatter's tea party, where . . . one proves something by proving what it is not."

The appeal board wondered why the licensing panel had failed to confront James Nelson's testimony. In a colloquy with Dignan, Farrar said he "almost fell out of [his] chair" when he read Nelson's "significant testimony." To ignore those remarks, as the ASLB had done, Farrar figured Nelson had to be "a graduate student or something," but, on the contrary, he had considerable professional stature. Dignan agreed that Nelson "financially is a good man" but maintained that the licensing board's apparent and unexplained rejection of Nelson's views did not constitute legal cause to revoke the construction permit.[20]

A majority of the appeal board grudgingly concurred. The ALAB found Nelson's remarks "particularly comprehensive and informed," and the licensing board's consideration of them inadequate. Harrison had obviously told different stories in different settings, and PSNH would probably experience difficulty in raising money for an investment that perhaps was insecure. But the New Hampshire PUC was likely to continue to approve rate increases as it had in the past. On that basis, the board found a "reasonable assurance" that PSNH could find the money to build Seabrook Station, which was all the NRC's regulations required.

Michael Farrar forcefully dissented. The majority's decision, he charged, implied that state regulators had the final word on finance, which essentially accepted Dignan's check-the-box test. "If we were not concerned with a nuclear facility," Farrar wrote, "I would be willing" to acquiesce. "But this is a nuclear power plant," he continued, "and that makes a difference."

> There is a need to avoid a situation in which financial pressures . . . become so pervasive as to influence the manner in which the plant is constructed. . . . We should not close our eyes to the likelihood that letting a financially strapped company go ahead with construction will inexorably result in decisions to do less testing, to use lower quality materials, to approve borderline workmanship, and the like. As I see it, an applicant must show that it will be able to obtain funds in ready enough fashion to avoid the likelihood that temporary shortages may compromise safety. The applicants have not shown this here. It invites disaster to overlook it.

The requirement that a corporation be financially qualified, in short, bore a clear relationship to safety, which was supposed to be the NRC's first concern. Farrar said he would have reversed the decision to permit construction.[21]

The full commission decided that the link between money and safety, which Farrar postulated, had not been demonstrated. "Close scrutiny of financial qualifications" had never before seemed necessary "for established utilities with substantial operating records"; the inquiry in the Seabrook case appeared "to have been the most searching examination . . . in the history of commercial power reactor licensing." The NRC was content to rely on the finding of the majority of the appeal board, which concluded that the necessary funds would be forthcoming and that the price PSNH might have to pay and the difficulty of obtaining them were not federal problems. "That being so," the commissioners concluded, "it is unnecessary for us to consider here the particular strengths and weaknesses of each witness' testimony."

Thus the NRC proclaimed itself off the hook. On the one hand, it explicitly disavowed Dignan's automatic test of financial qualification; on the other, the decision appeared simply to "check the box." In effect, the commission declared the testimony of all witnesses irrelevant, which enabled it to ignore not only Professor Nelson but also Robert Harrison's embarrassing contradictions. Finally, the commissioners could discover no connection between resources and safety, which had seemed obvious to one member of the appeal board and which furnished the logical base for the regulation itself. The decision may have been final and official, but it undermined the agency's own regulation and defied common sense.[22]

Although PSNH survived the NRC's scrutiny, the examination put a spotlight on the two financial faces of Robert Harrison and led to diminished public trust in the corporation and its leaders. This lack of confidence showed not only in higher interest charges but also in mounting doubt about the company's estimates of cost and demand for service, in cynicism about promises of safety and environmental protection, and in escalating doubt about any assertion of corporate expertise. Financial disclosure, in short, raised questions about the competence, reliability, and candor of management that could not be restricted to financial topics. The company itself sensed this loss of trust, which it sought to remedy with larger expenditures for "communications," or public relations. The company was challenged even in such previously friendly precincts as the NRC staff and the New Hampshire legislature. When PSNH repeated a year later Governor Thomson's estimate that licensing delay was costing $10 million each month—Harrison's figure was $300,000 per day—the staff dismissed the claim as "tenuous at best." Furthermore, the staff added, whatever the numbers, construction was at the company's risk and financial hardship was irrelevant. Dignan countered lamely that, regardless of who took the risk, the dollars were real and it was unfortunate to waste them. Martin Malsch simply reiterated the staff's flat judgment that PSNH's statement was "not a true reflection of absolute costs."[23]

New Hampshire legislators did not know what those costs might be, but their constituents, who heard terrifying rumors, began to grow restless. At about the same time the PUC approved the company's request to add the cost of construction to the rate base, the New Hampshire House of Representatives debated a bill that would have overturned that ruling and prohibited collection of CWIP. Like the PUC, legislators weighed the differing functions of customers and stockholders, the reputed saving of pay-as-you-go financing, the illusions of noncash earnings, and the dimension of the price increase that CWIP would impose. And the solidly Republican legislature, which ordinarily deferred to Governor Thomson, defied him and voted to prohibit CWIP. The governor's veto gave the utility a reprieve.

Management mistook that reprieve for a permanent stay. William Tallman, PSNH's president, seemed to a *Fortune* reporter to typify utility managers, engineers by training and temperament, who had "little feel for the world outside their control rooms." Tallman's contempt for politics, politicians, and public relations were ill concealed, but the fate of his corporation, in a sense, was tied to that of Governor Thomson, who probably had not even needed the urging he received from Tallman to veto the prohibition of CWIP.

But Hugh Gallen, Thomson's lightly regarded Democratic opponent, took the issue into the gubernatorial campaign of 1978. Gallen characterized CWIP as a disguised broad-based tax, which New Hampshire voters reflexively reject, and thereby staked a claim to the issue that had been the key to Thomson's political success. Gallen's effort to connect Thomson to CWIP, higher electricity costs, and PSNH was enhanced when directors of the company, which was supposedly struggling to pay its bills, voted in September to increase the dividend on common stock almost 13 percent. This distribution to PSNH's owners, increasing their return on investment above that of most other utilities, was also a bonanza for Hugh Gallen: "If the Lord himself had come down and said, Gallen, what do you need to win this election, I couldn't have asked for more."

A livid Thomson knew the dividend decision had a public-be-damned arrogance and tried to support Seabrook Station and CWIP without defending PSNH stockholders. But months later, when Thomson had lost the election and CWIP was likely to be prohibited, Tallman still defended the increased dividend, which he and Harrison had recommended without consulting anyone else. There had been discussion of political repercussions, Tallman remembered, but the board had believed Thomson would survive and, as one of the members put it, there was no ideal time for such a decision, so they might as well go ahead. Nor did the experience appear to have sharpened Tallman's capacity to assess the public result of corporate actions: "We were," he recalled, "winning the public relations battle [over CWIP] right up to the election."

That statement, among others, led the *Fortune* reporter to observe that "the criticism of company officials that seems hardest to refute is that they were just too optimistic." This sunny outlook had led them to miss every relevant estimate from the moment Seabrook Station was conceived, though Harrison blamed some of those errors on opponents. Thirty months after Nelson's testimony, for instance, when events had made him seem foresighted, Harrison still disparaged the economist with a fuzzy analogy: "If I say you're going to fall and break your leg, then I walk up behind you and hit you over the head with a stick, and break your leg, I can say I predicted the future." Harrison also admitted that the company's lenders had told him in 1977 that AFUDC would not suffice and "we had to get some real income before they gave us any more money," which seemed somehow more to the point than fantasies about academic economists bearing sticks. A pensive Tallman, looking back in 1979, thought there was "some question, in hindsight, . . . knowing everything we know now, and given our small size," whether the company would decide again to undertake construction of Seabrook Station. It was a rare, and revealing, admission. [24]

Hugh Gallen had barely had time to savor his unlikely victory before he had a visit from William Tallman. Gallen confirmed that he would not actively interfere with PSNH's effort to complete Seabrook Station but added that he would sign legislation barring collection of CWIP if the legislature passed it. That stand heightened uncertainty on Wall Street about the company and its revenues and caused Harrison to postpone a sale of common stock. CWIP had added perhaps $17–18 million to PSNH's revenues, and Harrison confidently expected to be able to replace that amount if the legislature acted, as it did in May 1979. "Utility rates or charges shall not in any manner be based on the cost of construction work in progress," the statute read; no cost could be included "until and not before said construction project is actually providing service to customers." [25]

A precise figure for the additional revenue PSNH derived from

CWIP in parts of two fiscal years is not easily calculated and probably unnecessary. Whatever the number, after 1979 the company's financial reports once more showed AFUDC, or noncash earnings, instead of the cash CWIP had provided. That situation forced management to find funds outside the company, not only to finance construction but also to pay dividends and interest. PSNH could not sell more first-mortgage bonds, so Harrison peddled subordinated debt, which was more speculative and bore a higher rate of interest. He mortgaged PSNH's share of nuclear fuel to, among others, two Swiss banks. He sold common stock when the market permitted and preferred stock when he could; some issues of the latter carried dividends of 15 percent or more. He loaded the company with high-priced debt, and he seriously diluted stockholders' equity. Again and again, he returned to the PUC for rate relief. And he had very little choice, unless construction stopped, which was the last thing Robert Harrison wanted.

Even before the CWIP prohibition passed, PSNH's directors had voted to reduce to 28 percent the company's share of the Seabrook project. In addition, PSNH would cut back its participation in two other nuclear plants under construction elsewhere in New England. The reduced expense—more than 40 percent of the company's projected expenditure at Seabrook alone—would more than compensate for the loss of CWIP revenue. But those arrangements, which would have postponed any cash outlay for Seabrook for a couple of years, fell apart. No one wanted the company's share in other nuclear plants, which therefore remained an expense rather than a source of revenue. Further, only a portion of the 22 percent of Seabrook PSNH put on the block could be sold, because regulatory agencies in other New England states discouraged additional participation while completion was uncertain and costs out of control. [26]

With its financial plans in ruins, PSNH had to improvise through 1979 and into 1980, while material and labor costs climbed and interest rates soared. On several occasions, including a filing with the Securities and Exchange Commission and a letter to their partners, corporate officials claimed that the sale of 22 percent of Seabrook Station was crucial to continued operation; each time, bankers or investors rescued the company before disaster struck. PSNH said it would run out of cash in April 1979, for instance, unless new sources of short-term credit materialized. In December, Harrison said the company might well be insolvent in six months unless the PUC granted immediate rate relief. The rhetoric always invoked catastrophe, which was no way to convince investors of the company's viability or to win the trust of customers and neighbors.[27]

Robert Backus mailed a batch of those statements off to Harold Denton, the NRC's director of licensing. What, asked Attorney Backus, does it take to demonstrate that a company is not financially qualified to undertake a nuclear project? Here was a situation where corporate officials told obviously contradictory tales to various audiences and could not carry out their financial plans. Foreign banks held a mortgage on nuclear fuel, which might have implications for national security. Payments to contractors had apparently been deferred, an action that stirred concern about quality control and safety. Several partners were trying to reduce their holdings, in some cases at the behest of regulators and in others out of ordinary fiscal prudence. Perhaps, Backus suggested, it was time to revoke the construction permit.[28]

Denton delayed his response while the NRC staff probed PSNH's financial health more thoroughly, he said, than regulations required. (The delay also permitted construction to continue without hindrance, which was precisely what Backus had meant to prevent.) In denying the request, Denton noted that PSNH had successfully marketed stocks and bonds and had enlarged its line of credit in the months since Backus's motion. "We know of no more convincing way of demonstrating a reasonable assurance that funds will be obtained," Denton wrote. And a "reasonable assurance" was all the commission required, even if "the necessary funds" only came "at a high cost"; the NRC did not demand "that an applicant's . . . financial outlook be rosy."[29]

Which, of course, it was not. About a month after Denton's decision, the PUC granted PSNH another increase in rates to meet another financial emergency. Once more the company had made the case that it would not be able "to sell long-term securities" or to "finance its construction or its day-to-day operations" without more revenue from its customers. By its own admission, as the PUC paraphrased PSNH's argument, the utility faced an "immediate and substantial financial disaster both as to the completion of Seabrook and the continuation of PSNH as a corporate entity." For his part, Harold Denton discounted that sort of talk: "dire statements by investor-owned, regulated utilities are common (especially during periods of high inflation and record-high costs of capital) in their quest for rate relief." He and the NRC staff relied less on "statements . . . before utility commissions" than on "the end result."[30]

The result, of course, was approval of the rate increase and, for good measure, a gratuitous endorsement of both Seabrook Station and PSNH's financial maneuvers. "If Massachusetts intervenors" had not interfered with PSNH's attempt to sell part of its share in Seabrook Station, the PUC said, the emergency would never have occurred. Since divestiture had been blocked, PSNH needed the access to financial markets that only greater revenue would create. Otherwise, construction at Seabrook would stop, and the PUC believed "that Seabrook is a valuable project." Thus the contention of consumer advocates that the rate increase concealed a charge for construction in progress, an allegation the PUC dismissed, seemed demonstrated to some extent in the commission's own prose. Armed with the additional revenue and an implied promise from the PUC to support Seabrook in the future, Robert Harrison went back to Wall Street.[31]

The perilous state of PSNH was notorious in the investment community. *Business Week* noted late in 1979 that most financial observers seemed to think that, "one way or another, Seabrook will be completed," though many wondered whether PSNH would be around at the finish. The best guess was that stronger partners would keep construction moving and save PSNH from bankruptcy. But a few months later, *Business Week*'s sources were more pessimistic about the utility's survival. In the weeks following the emergency rate increase, the company wanted to sell 2 million shares of common stock but cut the offering back to 1.5 million when the market price dropped below fifteen dollars, or about two-thirds of book value. The indicated dividend made the yield about 15 percent, which was about what some new PSNH bonds would return. But, *Business Week* reported, 90 percent of the income to pay those dividends consisted of construction allowances, so the company was not earning cash; rather, PSNH in effect borrowed from one group of investors to meet its obligations to others and would need "more than rate increases to solve its problems."[32]

And so it went for a couple of years until the PUC began to signal a limit to its tolerance and growing doubt about the data PSNH used to document its petitions. Not only did the utility tend "to overestimate its revenues and underestimate its costs," but the PUC wondered whether those estimates had been made in good faith: The "truth," the PUC feared, had "been at times manipulated, shrouded, ignored, unknown, and misconstrued." Those conditions did not support reasoned decision making and drove the commission to "assume that the company is actually worse off than their financial forecasts would indicate."[33]

By then, the NRC no longer cared. In March 1982 the commission changed its rules to eliminate review of the financial qualification of applicants for nuclear licenses. The commission decided, Denton explained to Backus, who was once more trying to pursue the issue, that the financial condition of a utility, "in and of itself," had no bearing on "the quality of construction and operation of a nuclear power plant." The NRC would not in the future consider financial matters as such but would deal with them only when they impinged directly on such regulatory concerns as health and safety. Whatever the dimension of PSNH's financial distress, Denton asserted, it had no discernible effect on quality, and it was, consequently, not the commission's concern. It had taken some years, but the NRC had decided to check Thomas Dignan's box. A federal court asked the commission to take a second look, which was perfunctory and affirmed the change. James Asselstine, in dissent, thought his colleagues' disposition of the issue would reinforce "the belief of many that this agency will go to any length to deny members of the public a fair opportunity to raise issues in our licensing proceedings and to have those issues fully and fairly litigated."[34]

Whatever the legal requirement, PSNH's need for cash remained. The company lurched from crisis to crisis, from Wall Street to the Public Utilities Commission, while Seabrook Station remained inoperable and outside the rate base. For a time, each new expedient must have held the promise of being the last. Reducing the labor force at the site, or slowing and then stopping work on the second reactor, or a major managerial reorganization, or suspending dividends on common and then preferred stock—all probably seemed at the time the elusive last decision that would enable PSNH to turn the final financial corner. None, of course, had that result. The weight of debt grew and interest compounded and rates went up and opposition multiplied. Professor James Nelson, and many others, could reasonably say that the whole script had been drearily predictable. Formal bankruptcy, which stared at PSNH several times in the years after Nelson's appearance in the licensing hearings, was temporarily forestalled. But the reality of insolvency had long since become evident in PSNH's financial statements.

The message David Freeman brought to the assembly of nuclear executives and energy officials from the Reagan administration in 1982 was not one they wanted to hear. Government, their party line ran, was the cause of delays and cost overruns that blocked an increase in the nation's nuclear generating capacity. But, said Freeman, "the view that all the economic problems would disappear if only the regulators got off the industry's back [was] a smokescreen," a misconception that threatened the industry more than excessive regulation. A director of the Tennessee Valley Authority, which had had considerable difficulty managing and enlarging nuclear capacity, Freeman spoke from experience when he charged that "the nuclear industry does not have a product that any utility in the United States can afford to buy." Operating companies could not accurately estimate costs or order design and drawings or guess at a construction plan that would resemble reality, Freeman said, because the fundamental safety issues had not been definitively resolved. He advocated the expansion of nuclear power, but he believed the industry needed a new technology based on a new reactor: It was time to stop fixing "the patches on our light water design."[35]

Public Service of New Hampshire could not afford to abandon the light water reactor, but the company's experience certainly illustrated Freeman's central point. PSNH's predictions of costs and dates of operation were notoriously inaccurate, and design problems frequently interfered with efficient construction. Freeman's remarks put those managerial deficiencies in an important context and indicated that they were not unique to the Seabrook project. But that context, however consoling, did not alter the discouraging circumstances the company faced.

The credibility gap on costs and calendar derived in part from the inherent difficulty of the task. Nuclear plants are inevitably complex, take years to build, and encounter unexpected technical impediments that require changes of design and additional dollars. The experience of the entire nuclear industry was not vast when PSNH undertook to build Seabrook Station, and PSNH itself had none. Yet guidance offered by the Atomic Energy Commission seemed calculated to entice utilities to undertake construction rather than to assist them with realistic budgeting. In 1971 the AEC said that capital costs should not exceed $150/kilowatt; by 1976, when construction commenced, the estimate had jumped to $1,200/kilowatt; in 1982, when PSNH was struggling to meet its obligations, the company believed it could complete the plant for $2,200/kilowatt; in 1985 *Forbes* reported that Seabrook, then almost complete, had already cost nearly $4,000/kilowatt. In retrospect, errors of that magnitude make management seem disingenuous or incompetent, and elements of both were unquestionably present. But the predictive task, it must be said, was extraordinarily difficult, and utility executives and nuclear bureaucrats were not the only Americans who failed accurately to foresee events of the 1970s that raised costs.

PSNH exacerbated the difficulty with multiple spokesmen, who delivered varying messages to various audiences. When describing the cooling water tunnels to environmentalists, for instance, PSNH tended to stress its great expenditure to protect the marsh; to establish the economic superiority of the Seabrook site over others that did not require tunnels, however, the expense could be made to seem less significant by the omission of interest charges and costs of delayed operation. In 1977, when the NRC was considering requiring cooling towers, a Seabrook construction executive estimated that towers would add almost $700 million to the eventual bill for the plant. Actual construction would cost less than $20 million; the other $660 million plus would pile up because of delay, purchase of power from other sources, and penalties because Seabrook was not available. Another estimate, less than three months later, had it that construction would amount to $50–60 million, and capital costs would be approximately ten times larger than that. The point, of course, was that towers were an absurdly uneconomic addition to the plant. But the estimates also seemed absurd.[36] Perhaps Robert Harrison's habit of not adjusting financial requirements to changes in design was based on his review of calculations like those for the towers.

Since the cost of components was susceptible to enormous variation, the total cost of the project could not be a simple exercise in addition. The series of "official" estimates, which Seabrook owners issued in response to demands of regulatory agencies, seem with hindsight both random and ridiculous. Totals rarely resembled numbers intervenors used or those independent consultants reached. As early as 1975, for example, Guy Chichester of the Seacoast Anti-Pollution League (SAPL) said the plant would cost about $2.6 billion; the figure the company released in reply was exactly half as large, though for its own internal purposes PSNH was already using a number several hundred million dollars greater. And the estimate kept climbing at irregular intervals, to $2 billion in December 1976, to $2.6 billion in January 1979, to over $3 billion in April 1980, and then to more than $5 billion in November 1982, by which time the difficulty of obtaining an accurate estimate was only a symptom of other, and more serious, managerial shortcomings.[37]

A crucial variable in cost calculation was the date assumed for commercial operation, which in turn guided other assumptions about the price of goods, services, and money. Management decided to use aggressive scheduling as a tool to keep the pressure on contractors and employees, or at least that was the explanation one consultant reached after looking at the unrealistic calendar. But the result of that decision, one critic observed, was a timetable that could not be kept, morale problems, and multiple inefficiencies because of the interdependence of schedules for construction, purchasing, and finance. Although PSNH knew the project was more than a year behind even before the NRC began formal hearings, the company was reluctant to acknowledge any slippage. Changes thereafter never quite caught up with conditions at the site: When the state siting committee finished deliberations in 1973, Unit 1 was to be on line late in 1979; when construction actually began in 1976, the date had been moved back twelve months; in 1979, after Gallen's election as governor, the company thought operation would start in the spring of 1983; in the spring of 1984, the predicted date was July 1986. And so on. There is no evidence that fictitious deadlines stimulated speed or productivity, if that was their object. Tardy segments of the work force, indeed, simply blamed lagging performance on a schedule everyone knew was unrealistic. Management, moreover, seemed out of touch with construction reality.[38]

From the outset, the managerial structure of the project perplexed the NRC. Clearly, PSNH as the majority owner was to act for the other eight or ten utilities that at various times held shares in Seabrook Station. But the relationship among PSNH, Yankee Atomic Energy Company (YAEC), which was to supervise construction and engineering, United Engineers and Constructors (UE&C), which was to design and build the plant, and various subcontractors, who would do the work, was never spelled out to the NRC's entire satisfaction before 1984. Even while allowing construction to begin in 1976, the licensing board noted that PSNH did not have appropriate management systems and appeared "to rely more heavily on the judgment and response of individual operators than seems warranted." The "corporate base in direct support of [the] nuclear program" seemed to the ASLB "only . . . modest," and PSNH would have to depend "heavily on its contractors and consultants for . . . technical support." The participation of Yankee Atomic, which had managed the construction and operation of other New England nuclear plants, seemed essential to provide the expertise PSNH lacked.[39]

Although William Tallman assured federal officials as early as 1974 that PSNH and Yankee were working together without friction, the division of responsibility continued to baffle people outside the two organizations until a major corporate reshuffling in 1984. As late as 1983, a member of the Advisory Committee on Reactor Safeguards told David Merrill, the executive vice-president of PSNH, that he had never before seen so many colors required to map an organization. People appeared to migrate from payroll to payroll and might not recall who their employer was if they were called in the middle of the night, mused another member of the panel. The Management Analysis Corporation (MAC), hired by PSNH during the managerial crisis in the early 1980s, pointed immediately to the ill-defined relationship between PSNH and Yankee. On the one hand, the consultant reported, some thought of Yankee as the "eyes and ears of the owners," an acute observer to keep PSNH informed; others believed Yankee should act for the owners and make construction and design decisions. No one, MAC asserted, seemed entirely sure which model had been adopted, and the consequent confusion created layers of hierarchy, divided responsibility, and wasted time and money.[40]

Managerial control was no trivial matter at the NRC. The report of the President's Commission on Three Mile Island, one volume of which focused on management, noted that better control of information by the operating utility might have averted the accident in Pennsylvania. But Metropolitan Edison, which ran the plant at Three Mile Island (TMI), like PSNH, had little firsthand nuclear experience and relied on contractors, vendors, and other corporate partners for expert assistance, of which there was, in a sense, too much. Amid all the paper that flooded Metropolitan Edison, there had been data that should have enabled operators to control the incident at Three Mile Island before it became a crisis. But those data never reached the proper hands with the proper emphasis. Managers managed; operators operated; papers were shuffled, initialed, and filed, and the memorandum that called attention to the very situation that disrupted TMI "fell through the cracks."[41]

The managerial problem at Seabrook was compounded by a strategy that intentionally minimized PSNH's day-to-day supervision and was probably intended to control administrative costs. The company did not add staff with nuclear experience and assigned executives specific tasks at Seabrook in addition to those associated with the utility's traditional business. David Merrill, for instance, had important operational responsibilities with PSNH and was also the utility's chief executive for Seabrook. Merrill's assistance was to come in large part from consultants and contractors who were not on PSNH's payroll. The geographic dispersion of management suggested that direction of thousands of workers at the site could be remote: Merrill's office was in Manchester, about an hour from Seabrook; Wendell Johnson, who was the Yankee executive for the project, had other duties to perform in his office in

Framingham, Massachusetts, which was also about an hour away; most of the engineers and executives of UE&C worked in Philadelphia.

Management consultants highlighted that situation in reports that pointed to "inadequate management presence" and "failure" to designate "a single strong manager who was clearly in charge." James B. Baker, president of a group of municipal utilities that together owned about 15 percent of the project, told Robert Harrison in 1983 that "managerial control and accountability" sufficient to complete and operate the plant seemed absent. Baker believed PSNH needed "qualified managerial assistance" immediately because "tight and detailed owner-management control" was essential "for the successful completion of so complex, expensive, and critical a project." Yankee, in Baker's view, would not do. [42]

Allegations of managerial inefficiency tended to blame organizations and systems rather than people. Whistle-blowing construction workers, on the other hand, whose affidavits were collected in an effort to persuade the NRC to undertake a searching reinspection of the plant, often pointed to the inadequacy of individuals. One worker claimed, for instance, that a person in charge of the central blueprint office "had no idea how to . . . update changes to make sure the crews had the most current prints." Others working in that office, he added, could not even read drawings, let alone update them. A second worker confirmed that prints were often incorrect: "we would go to this area and there would be various other systems already there," a situation that required consultation and occasional redesign and rebuilding. A foreman said that drawings sometimes differed completely from the situation he and his crew encountered, a discovery that led to delays—often for weeks—while engineers investigated. As a result, "some crews had no work to do for several days at a time." A laborer, accompanied by three electrical engineers and joined by a fourth, dug for several hours in a futile search for a manhole that was not where the blueprints said it was. Marijuana use in the blueprint center was constant, Peter Hanson charged, and sometimes "really stoned" workers there tossed "revised blueprints into the shredder" to save the effort of finding and replacing outdated drawings.

Blueprint clerks had no monopoly on malfeasance. A welding inspector was convicted of falsifying records, and several batches of apparently inadequate welds had to be reinspected and in some cases repeated where they were not already encased in concrete. The PSNH site manager complained in 1980 that "the quality of welding is poor and seems to be worsening." Whistle-blowing welders said that they could not maintain their morale in the face of insistent demands for overtime and capricious inspectors. Nor could they sustain pride in their craft when many of their fellow workers had little skill and no experience. Because of the shortage of able welders, the union established a school to certify journeymen after about three weeks of instruction; many of the graduates, one disgruntled welder said, were friends and family of those with influence at the plant or in the union. This sort of favoritism also affected inspectors' reports, and good welds often had to be repaired whereas flawed ones passed. Many pipes contained discarded tools and construction debris that tired or distracted welders had dropped and forgotten, according to one of them, who finally quit after four years on the job. "This job sucks" was his inelegant conclusion, and that "was the overwhelming attitude throughout the plant."[43]

Yet, considering the thousands of tradesmen who worked at Seabrook in the years the plant was under construction, the number of disaffected workers willing to go on the record was very small. To be sure, the work force dispersed as construction wound down and could not be expected to respond to appeals from those who sought to substantiate rumors of shoddy or unsafe construction. Further, there is evidence that unions and fellow workers intimidated those with misgivings about craftsmanship or design. When allegations did surface, neither management nor the NRC had much incentive to pursue them with vigor, though NRC inspectors did return to Seabrook to look into, and dismiss, charges made by the Employee's Legal Project, which had gathered statements from whistle-blowing construction workers.

The credibility of those whistle blowers does not rest on their number, which was small, or on the anecdotes themselves, many of which are not susceptible to independent verification. But the reports of consultants to management, which used different vocabulary to describe some of the same phenomena, corroborate the narratives in part. Seeking to explain why so many people and so many dollars produced so little result, for example, consultants pointed to the failure of subcontractors to design pipe hangers on schedule; whistle blowers identified incompetent, or lazy, or intoxicated people who shredded blueprints. Consultants blamed Pullman–Higgins, the piping subcontractor, for failing to prevent welding deficiencies; whistle blowers described inept or untrained or careless welders. Consultants, and to some extent the NRC, sought general and managerial solutions to problems for which whistle blowers offered personal explanations. The perspectives differed, but the accounts often complemented one another.

In particular, management consultants repeatedly criticized Pullman–Higgins for failure to adhere to the schedule for design and installation of piping. Much of the work assigned to Pullman–Higgins was on the "critical path" and had to be completed before other tasks could start, but, in spite of increased payroll, the schedule continued to slip. In November 1982 piping installation had to be halted to permit Pullman–Higgins's engineers to bring designs up to date. Although the backlog shrank, the NRC noted a few months later that "only marginally acceptable results were achieved." In 1983 PSNH told its partners that piping installation was nine months behind schedule. One consultant, engaged by the Vermont Public Service Commission, concluded that, in the final analysis, PSNH and Yankee as well as Pullman–Higgins were responsible:

> the inability to resolve the problems with Pullman–Higgins is a reflection of the lack of effective management of the construction process by PSNH and . . . Yankee. This ineffectiveness allowed chronic problems with this subcontractor . . . to continue for a period of *years*.

Even a report prepared for minority partners of the enterprise, which tended to praise management's effort, reproved Pullman–Higgins for the perpetual backlog of incomplete drawings, though the Challenge Consultants believed the resulting delay amounted to weeks and was not critical.[44] To resolve the issue, late in 1983 UE&C undertook to design what remained of the piping for Seabrook Station.

By that time there were serious questions about the performance of UE&C as well. As early as 1981 an internal report by a task force on planning and scheduling had complained that "design and construction [were] proceeding on a hand-to-mouth, crisis basis." In particular, the sluggish system for making authorized design changes as construction continued led to unnecessary expense and confused tradesmen who encountered circumstances at variance with plans. In 1982 about 4,600 design changes were not documented; a year later, the number was more than 18,000, and the situation, the Vermont consultants said, was "completely out of control."

> We conclude the UE&C's inability to control the design change process was a direct result of the ineffective management structure at Seabrook prior to the 1984 reorganization.

That reorganization began with a report in June 1983 from the Management Analysis Corporation. Before the end of the year, David Merrill "decided" to take "early retirement"; William Prince, a member of the MAC team, who had recommended that PSNH itself actively supervise construction, temporarily assumed Merrill's responsibility for Seabrook Station. In January 1984, UE&C produced a new schedule and new cost estimates, which Prince said were unreliable. Two months later, Harrison reported to the joint owners that he had forcefully expressed their dissatisfaction to UE&C; he also reported that he had ordered legal research to see whether the owners might seek a financial settlement from UE&C and other contractors on the project.[45]

John Sununu, the governor of New Hampshire and as vigorous an advocate of Seabrook Station as Governor Thomson had been, was not interested in assigning culpability, but his prescription for PSNH's Seabrook ills implied that previous managerial people and systems had been inadequate. Sununu urged the company to recruit a team of nuclear professionals able to take control of construction, testing, and then operation of Seabrook Station. However capable, utility executives with experience in coal or oil plants would not suffice, the governor said; he thought PSNH must recruit "a few dozen" senior officials in order to create "a critical mass of management capability." PSNH had delegated too much responsibility to subcontractors, Sununu told Harrison, and needed to develop its own capacity to manage the facility. He suggested also that PSNH acknowledge the de facto abandonment of the second reactor and concentrate on completing the first.

The governor's intervention was both public and extraordinary, not just because it foreshadowed the company's eventual actions. Sununu's support of the controversial project, and that of his administration, had never been in doubt. But by this action, he also became in effect a consultant to management, providing advice that the company might reasonably regard as the price of his continued blessing. In any case, PSNH embraced the governor's advice; Harrison said the company had been seeking a nuclear management team since Merrill's retirement. About ten days later, the search concluded with the introduction of William Derrickson, PSNH's new vice-president for nuclear engineering, who had a deserved reputation for on-time, on-budget, active administration of precisely the kind Sununu had recommended. Thus the governor appeared to be a wise and vigorous public official urging on one of the state's crucial corporations a course so sensible it had been adopted instantly. And PSNH, which had previously seemed managerially stubborn, now displayed imagination, flexibility, and enterprise in recognizing and acting promptly on wise counsel. William Derrickson, the visible result of this happy collaboration, could begin his task with all the momentum that an unusual union of the state and private enterprise might provide.

Yet the negotiations leading to Derrickson's appointment, and the decisions of the PSNH board that led to them, must have been well advanced before the *Manchester Union-Leader* broke the story of Sununu's intervention. The paper had long lent its news columns, as well as its editorial page, to the support of PSNH, Seabrook Station, and the governor; the story derived not from journalistic enterprise but from Harrison's release of a letter from Sununu that purported to summarize conversations the two had held in the recent past. Those conversations may well have taken place, and Sununu may well have given Harrison good advice. But the story itself seemed contrived for both corporate and political purposes.[46]

William Derrickson arrived in Seabrook with all the subtlety of a hurricane. Workers at the plant sensed a driving purpose, in part no doubt because the new boss walked around firing people. Within weeks, a thousand employees had departed, and a new emphasis on productivity had been effectively communicated to all at the site.[47] But this purposefulness was not evident in the estimates of the Management Analysis Corporation, released about two weeks after Derrickson's appointment, that Seabrook Station would cost about $9 billion—up from the most recent official figure of $5.2 billion—and would be delayed nineteen months beyond the previously announced schedule. PSNH released the consultant's predictions, together with a statement disputing them, claiming they did not take account of improvements that would be achieved as a result of reorganization. But the numbers dismayed bankers on whom Robert Harrison called in search of additional credit: "'Don't ask,'" they said, Harrison remembered, "and that is pretty much a direct quote."[48]

The inability to secure new loans forced administrative reorganization in addition to Derrickson's appointment. Under pressure from regulatory authorities, utilities in Maine and Vermont put their shares of the project up for sale. Other partners sought to curb PSNH's control through revisions of the management agreement and advocated abandonment of the second reactor. In May 1984, PSNH could not pay its ordinary bills, to say nothing of the extraordinary costs of construction, which were exceeding $300,000 each day. The company missed a $5 million payment to its short-term creditors, an installment on the purchase of nuclear fuel, and part of its franchise tax to the state. Without operating funds and without credit, PSNH was within weeks of becoming the first utility to declare bankruptcy since the Great Depression.

The formal step, for the moment, was avoided. Dividends on common and preferred stock were omitted. A debt to UE&C of more than $20 million was converted to a loan. A new corporate division—ultimately New Hampshire Yankee—was created to complete and operate Seabrook Station, thereby terminating PSNH's control of the project. Expenditure on the second reactor stopped; indeed, for some months, all construction stopped. Wages and salaries were frozen, and assets, where possible, were sold to raise cash. And Robert Hildreth, of Merrill Lynch, who shaped this scheme, then sold $90 million in short-term bonds that temporarily kept PSNH afloat; the 20 percent interest charge was actually lower than the average 24.4 percent PSNH had paid on its short-term debt the year before. Then the company went back to the PUC for approval of $450 million in long-term financing that would see the company through 1986, when the first reactor, dubbed "Seabrook I," would presumably be licensed, operating, and added to the rate base.[49]

The PUC cooperated, though final approval required the intervention of Governor Sununu and a trip to the state's Supreme Court. A member of the PUC, who had made several speeches endorsing the company's strategy before the matter formally came before him, had to be disqualified; Sununu then appointed an officer in a nuclear trade association as a temporary replacement, who made the expected decision. The NRC cooperated by dismissing arguments from NECNP and SAPL that the new entity could not inherit PSNH's construction permit without formal hearings. And work recommenced at the site, with results that seemed to NRC inspectors to indicate that many of the previous deficiencies had been cured.[50] The managerial reorganization, however tardy and however expensive, permitted completion of Seabrook I.

And, in spite of a record that a dissenting justice of the New Hampshire Supreme Court called "abysmal," corporate officials, with a couple of exceptions, retained control of the state's largest utility. The continuity of those executives, given the breadth of the corporate disaster over which they presided and for which in some measure they were responsible, belies much of the folklore of capitalistic accountability. In 1976, when construction began at Seabrook, PSNH was celebrating its fiftieth anniversary. Robert Harrison, then in his twentieth year with the firm, was the junior member of top management in point of service; William Tallman and several others had joined PSNH a decade before Harrison. A generational turnover coincided with the financial crisis of the early 1980s, when Tallman relinquished operating responsibilities to Harrison and David Merrill and several others retired. With the exception of William Derrickson and a few colleagues who joined in 1984, Harrison's new managerial team consisted of old company hands. In a sense, then, the stockholders relied on the very folks who had steered the company toward bankruptcy to find a route around it.

The longevity of management mirrors that of the board of directors, which also changed very little until the early 1980s and which has stabilized since. Four of the directors, including Tallman, have been reelected annually since the Seabrook project was conceived; the service of several others, including Harrison, spanned most of the years until 1983–4. Although preferred stockholders have been entitled to elect a majority of the directors since 1985, the fact has had no discernible effect on the composition of the group. Among the seven nominees of preferred stockholders in 1987, for instance, were two members of management and two other directors whose tenure antedated the Seabrook project.

If continuity in the executive suite and on the board can be interpreted as a sign of stockholders' approval, the owners of PSNH shares must have been a complacent lot indeed. Until dividends ceased in 1984, the yield on current market price was reasonably attractive. But the current market price was apt to be a fraction of what any long-term investor had paid for the stock, which had sold for more than thirty dollars a share before PSNH decided to undertake construction of Seabrook Station; at the bottom, in 1984 and 1987, the market price of the same share was less than four dollars. In part, the decline reflected a dilution of ownership through several issues of common stock in intervening years; in part, of course, market value simply revealed the investing community's assessment of the company's dismal financial outlook.

Neither management nor the directors had a large personal interest in the corporation. Only eight of the directors nominated in 1987 owned as many as 1,000 common shares, an investment of less than $10,000. Of that group, three were members or former members of management, and three others had added to holdings since the suspension of dividends in 1984 and must, therefore, have been betting on appreciation rather than a predictable return on their money. With the exception of management directors, it seems likely that every member of the board but one, or perhaps two, collected directors' fees from the company in 1986 that exceeded the total value of their holdings in the company's stock. The board, in other words, had little collective long-term stake in the enterprise.[51]

Those in charge of American businesses, including individuals in PSNH's headquarters in Manchester, have occasionally relieved irritation and frustration by denouncing the bureaucratic evils of government. The comparison, implied or explicit, is the dog-eat-dog competition of free American enterprise, where salaries mirror merit, jobs are always insecure, and incentive produces outstanding results. In fact, as the persistence in PSNH's executive suite demonstrates, corporate managers are among the most entrenched and least accountable of the nation's bureaucrats, even without "poison pills," "golden parachutes," and other exotic protective devices. Dissatisfied stockholders, more concerned with preservation of capital than with administration of a company, tend to sell their shares rather than mount a futile campaign to replace managerial employees. The most forceful voices for reorganization at Seabrook came from politicians and from managers of PSNH's partners, not from the stockholders who owned the company.

Yet the decision to construct, and then to complete, Seabrook Station was manifestly contrary to the interest of both individuals and groups that in theory had the power to change those decisions. Neither stockholders nor PSNH's management would have voluntarily tied the company's fate to Seabrook Station had they been able to foresee the future. Wall Street financiers may have made a short-run profit on PSNH securities, but they did no service to their own reputations or for the long-term investors upon whose good will the securities business depends. The NRC, for all the bureaucratic imperative to expand production of nuclear power, can hardly argue that the example at Seabrook will inspire imitation. The utility's customers have involuntarily traded a staid and reliable supplier of electricity, and reasonable rates, for the uncertain prospect of more supply and the absolute certainty of unreasonable rates. An idle expenditure of more than $5 billion is no source of pride to any part of the nuclear industry—from the laborer at the site to the board rooms of Westinghouse and General Electric. Seabrook Station, to put it briefly, was a mistake that everyone concerned would like to believe just happened. Since no one is accountable, there is no self-evident address for the bill. And since no one wants to pay, everyone will.

Managerial bungling in the nuclear industry was not restricted to Seabrook Station. In a major article in *Forbes*, James Cook called the development of nuclear power "a disaster on a monumental scale." Expenditure, Cook pointed out, exceeded that for the space program or the nation's war in Vietnam, and "only the blind, or the biased, can now think that most of the money has been well spent." "The truth is," Cook charged, "that nuclear power was killed, not by its enemies, but by its friends." His indictment included the federal government and the NRC, manufacturers, utility managers, and state regulators—all of whom apparently had ignored the cost of the technology and thereby betrayed both investors and consumers. But Cook's central target was management, which was simply inadequate to the task.

PSNH was not in this respect worse than other builders of nuclear plants. Bad welding plagued every project, and many others had even greater difficulty with document and design control than did Seabrook. At least the reactors at Seabrook were not installed backward, as they had been at a plant in California. And at least the NRC had not approved a reactor design the utility was not planning to use, as had occurred at a plant in Louisiana. And at least there were not shelves full of reports of inspections on welds that never existed, which was the case at a plant in Kansas. And at least the reactor supports were properly placed, not off by a full forty-five degrees, as had happened at a plant in Texas. Thus the protests of PSNH and New Hampshire Yankee that Seabrook Station was well built had comparative validity.

But management's responsibility, as one Wall Street analyst noted, went beyond the duty to insure first-rate construction. The failure to understand that construction, of whatever quality, exceeded a corporation's means also constitutes mismanagement, and "in terms of its financial viability," the analyst continued, "Public Service of New Hampshire should never have built Seabrook." That fundamental error was not the fault of contractors or labor unions or regulations or environmentalists or obstructive politicians. A company that ties up "close to 80% of its capital in a construction project," especially one that by law cannot produce income until complete, ought not to blame others for wounds that were self-inflicted. [52]

5 | Emergency Planning

"Why make umbrellas if it's not going to rain?"

The Advisory Committee on Reactor Safeguards (ACRS) was approaching the end of its first discussion of Seabrook Station in 1974. James MacDonald, an official of Yankee Atomic Energy Company, spoke rather vaguely about arrangements for possible nuclear emergencies. Members of the committee knew firsthand about one aspect of the problem; they were meeting near the region's beaches in the summer, and traffic, tourists, and crowds were inescapable. Nunzio Palladino, a member of the ACRS and later chairman of the Nuclear Regulatory Commission (NRC), asked what changes in the region's roads were contemplated "in order to assure a good evacuation plan." MacDonald replied that the state had abandoned efforts to improve highways to the beaches because the surrounding marsh posed too many obstacles; "we haven't called for any roadway modifications," MacDonald said, "and don't intend to."

Palladino was not satisfied. If existing highways were inadequate, the difficulty of reconstruction should not prevent improvement. Palladino kept probing MacDonald's confidence until the latter conceded that no one had seriously investigated the matter. So, Palladino asked, would someone now undertake that investigation? MacDonald thought not; he said the company could probably write effective emergency plans based on existing roads. Palladino wondered aloud where that assurance came from, since no one had even looked at the question. "I wouldn't belittle the road network as a problem," Palladino concluded. "It looks like a problem to me."

MacDonald's promise of further study did not take him off the hook. William Kerr wanted to know the responsibilities of various state officials in the event of a nuclear emergency. MacDonald used words like "coordinates" and "communicates" that did not seem to Kerr to connote decisions and actions. He asked MacDonald whether the sponsors of the plant had calculated the probability of an evacuation. "Not numerically," MacDonald replied; "we never expect to evacuate or use our emergency plans." That response, Kerr said, implied that the probability was zero, which he did not accept and which he suspected the utility did not believe either. He tried again: "Do you have any estimate of what the probability would be?" "No numerical value," MacDonald responded.

Other members of the ACRS wanted to know about prevailing winds, whether roads and winds tended to go in the same direction, and about access to transportation. None of the questions received a definitive answer, and Palladino summed up: Emergency planning, he remarked in the summer of 1974, "may be one of the more contentious points on this site." The ACRS made the point formally in a letter to the Atomic Energy Commission (AEC) late in the year.

> Because of the proximity of the Seabrook station to the beaches on the coast and because of the nature of the road networks serving the beaches, the applicant has given early attention to the problem of evacuation. The committee believes, however, that further attention needs to be given to evacuation of residents and transients in the vicinity even though they may be outside the LPZ. [1]

Federal regulations at the time did not require emergency planning beyond the LPZ, or low population zone, an indefinite area that varied with the characteristics of the plant in question and the population of the surrounding area. Regulations required the LPZ to be susceptible to evacuation before inhabitants would receive excessive radiation as the result of a "design-basis" accident, a provision that lacked precision on several counts. No federal rule specified a "safe" dose of radiation, although there were limits that could not be exceeded in the LPZ. A design-basis accident, according to regulation, was one in which safety devices at the plant kept escaping radiation within acceptable limits; more serious accidents, which might result in greater radioactive releases, were, by definition, "incredible" and therefore need not be anticipated with emergency plans. The area of the LPZ, and the number of people who might live within it, depended on roads and weather and hardware and other site-specific circumstances such as those about which members of the Advisory Committee on Reactor Safeguards inquired.

The beaches—and the thousands of people on them in the summer—were notoriously the site-specific circumstance that made the Seabrook location problematic. Public Service Company of New Hampshire (PSNH) proposed to build the plant less than two miles from popular beaches in Seabrook and Hampton; a ten-mile circle enclosed much of New Hampshire's coastline and several miles of Massachusetts' as well. That larger area included state parks, trailer parks, and amusement parks, motels, movie theaters, mansions, campgrounds, and crowded sand. And that fact troubled the regulatory staff from the outset, because the population, at least during good summer weather, exceeded an informal agency limit of 400 people per square mile in the immediate vicinity of a nuclear plant. In a 1973 memorandum, which articulated the doubts of others members of the staff, the assistant director for site safety recommended that the utility "identify suitable alternate sites or definitely demonstrate that the lower population sites should not be selected." Intervenors too pointed to evacuation planning as the place "where we really feel the emperor has no clothes." The state of New Hampshire, through an assistant attorney general, wondered what the AEC would consider a safe period of time for evacuation of the beach, a query the agency deflected to PSNH, which, in turn, promised an evacuation plan that would answer all questions within eight weeks, by June 1974.[2] The promise was a trifle premature.

For the next few years, in the legal jousting surrounding the issuance of a construction permit in July 1976, PSNH argued that it had no responsibility to plan for any area beyond the boundary of the LPZ, which had been drawn intentionally to exclude the beaches. The Atomic Safety and Licensing Board (ASLB) agreed that PSNH had no legal obligation outside the LPZ and ruled against intervenors, who sought to extend the scope of emergency planning with the support of the NRC staff. Further, the ASLB decided that "the realistic consequences of design basis accidents" would not require "evacuation of persons beyond the . . . LPZ." Because of the plant's safety features, indeed, the board believed evacuation of people even from within the LPZ "rather improbable."[3]

Since no plans were necessary, the ASLB did not need to estimate whether a safe and speedy evacuation of the region could be achieved. The issue was not thoroughly aired during hearings in the summer of 1975 that led to the construction permit, though the captain of the New Hampshire State Police testified that the utility's conjectures about congestion and the time required to clear the beaches seemed to him unduly optimistic. The board simply recommended that PSNH and the state "cooperate" to complete "such plans . . . as the state may wish to employ in the unlikely event of an accident." Evacuation was not, in other words, a federal problem.[4]

Donald Stever, the assistant attorney general who represented the state in the licensing hearings, was dumbfounded. Ordinarily more reserved than counsel for other intervenors, Stever was sharply critical of the ASLB's "crabbed and inadequate findings," which he said would permit siting a reactor in New York's Central Park. (Thomas Dignan, PSNH's counsel, later remarked that in fact he could make the case for that location.) The ASLB, Stever charged, had ignored evidence of past population growth on the seacoast and had failed to look at projections beyond 1980, which showed continuing increases. Worse, the board had accepted ingenious, but unprecedented, statistical devices that minimized the effect of a possible accident, a calculation Stever called "so misleading as to constitute material misrepresentation of fact." In "an attempt to avoid reality," the staff had invented a statistical "artifice" that enabled the board "to substitute hastily contrived, approximating guesses" for the caution regulations required. The result permitted everyone to forget "about protecting the public on the beach."

The siting decision was only part of the error; failure to analyze the feasibility of evacuation compounded it. According to Stever, the board had understated the population at risk and then ignored expert testimony about the time required to relocate even those people. He did not want the task pushed off on the state, and he did not want the board "to alter . . . the regulations to suit a site." Rather, Stever wrote, "the site must be altered, . . . or rejected." He refused to say that a nuclear plant could never be built at Seabrook, but he did insist that federal authorities, and the utility, adhere to regulations that seemed to him essential for the protection of New Hampshire residents. Those regulations and that protection, Donald Stever feared, had been interpreted away.[5]

Members of the staff of the Nuclear Regulatory Commission separated the question of evacuation from their assessment of the site. They had previously agreed that a plant could be built at Seabrook; that was no longer at issue. But the NRC's basic responsibility under the Atomic Energy Act to protect the health and safety of the public appeared to require provision for people beyond the LPZ. For Michael Grainey of the NRC staff, the matter was "that simple." But John Buck, a member of the appeal board, had another view of simplicity; lacking a regulation to accomplish its purpose, the staff had simply invented one. Subsequent discussion demonstrated the legal precariousness of the staff's argument, which did indeed contradict previous rulings of the commission, as Dignan repeatedly pointed out. Counsel for the NRC staff did not often disagree with counsel for PSNH. That rare circumstance, combined with the evident vigor with which Grainey tried to make a weak case, suggests the staff's unusual concern about emergency planning for the region surrounding Seabrook Station.

Concern, however, was not law, about which, as Thomas Dignan remarked, there was little doubt. On six previous occasions, federal regulators had found no requirement to plan for people outside the LPZ. As a practical matter, if the LPZ was not the boundary, there was none, Dignan asserted, and licensing authorities "are going to kill 10 days in every future hearing" trying to determine the area for which emergency plans were required. He also saw no reason for his clients to spend "a pile of money . . . because the staff has a new ratchet" without any regulatory basis. The absence of evacuation procedures, moreover, did not leave people unprotected. Their safety, he said, rested instead on the millions of dollars in concrete, steel, and design that his clients were investing in the plant, hardware and technology that protected the public more surely than fallible emergency plans. The appeal board may not have accepted all of Dignan's argument, but the judges did agree with his reading of the law. If the staff wanted to require emergency planning beyond the LPZ, the panel ruled, the Nuclear Regulatory Commission would have to write a new rule, which occurred after the accident at Three Mile Island.[6]

Before Three Mile Island, a majority of the American public apparently believed the reiterated assurances of public officials that nuclear generation of electricity rested on a proven technology that was essentially safe.[7] Proponents of nuclear power maintained that fission posed fewer hazards to public health than those notoriously associated with the mining and combustion of coal, the major source of the nation's electrical power. One nuclear physicist and former chairman of the AEC wrote in 1971 that the probability of a catastrophic nuclear accident was about equal to the danger to baseball spectators at Shea Stadium from the simultaneous collision of every airplane in the vicinity of New York City during a Sunday afternoon doubleheader. Dixy Lee Ray, another chairman of the AEC, claimed that nuclear power was safer than eating because 300 people annually choked to death on food, whereas no one had perished in a nuclear accident.

The confidence of the commissioners pervaded the agency. Lack of a provision for compulsory state and local emergency plans before 1980 was no inadvertent omission but a deliberate decision to avoid dilution of the NRC's authority by involving governors and city or county officials. The President's Commission on Three Mile Island discovered fear within the agency that discussion of emergency planning might alarm the public and inhibit the industry's development. Those who raised the topic found their memoranda "disappeared into the sand like a glass of water in the Sahara," as did early concerns about Seabrook's beaches. Emergency plans, as Lee Gossick of the NRC staff articulated the agency's line, provided only "an added margin of protection" in circumstances where "we believe that an adequate measure of safety *already* exists" because of engineering and design. The prevailing attitude in the NRC before Three Mile Island, as summarized by the president's commission, was "Why make umbrellas if it's not going to rain?"[8]

But a substantial fraction of the public—perhaps a third nationally and a larger percentage in coastal New Hampshire—was not so sure. Maybe nuclear power was safer than coal, one skeptic remarked, "but the fact remains that we did not dump a ton of coal on Hiroshima." Although accidents at the Fermi reactor near Detroit and at Browns Ferry in Alabama had been controlled, those incidents reinforced the view that nuclear power was not without risk. As one reactor engineer put it, "if people want to turn the lights on, they are going to have to expect to lose a reactor now and then." The AEC's own 1957 safety study, which asserted the improbability of a "worst-case" accident, nevertheless came up with appalling statistics: 3,400 dead; 43,000 cases of radiation illness; 182,000 people with increased susceptibility to cancer; $7 billion (1957 dollars) in property damage; residential restriction on an area roughly the size of Ohio, Indiana, and Illinois combined. An attempt to review the problem in 1965, when reactors were larger, brought results the AEC found even less palatable, and the revision was not publicly released.[9]

Nor did the revision affect the public posture of the AEC, which steadfastly maintained that the technology was safe and lingering public apprehension irrational. Yet much of that technology had not been experimentally verified, an omission in sharp contrast to the government's frequent assertions that nuclear weapons must be constantly tested. This absence of a compelling, scientifically respectable demonstration of nuclear safety threatened to interfere with the expansion of an industry that utilities, politicians, and the president of the United States thought urgent in the early 1970s.

The *Reactor Safety Study*, released in 1974 and billed as an independent assessment of the hazards of nuclear power plants, was supposed to fill that gap. Prepared under the direction of Professor Norman Rasmussen of the Massachusetts Institute of Technology, the report was variously called the Rasmussen Report, the MIT Report, and WASH-1400. But it was not an independent study, and it was not prepared at MIT. Much of the data for the two plants studied came from Westinghouse and General Electric, and some of the computer analysis was performed by vendors of nuclear equipment. Professor Rasmussen conceded that he "had only a vague idea of how to carry out such a study" when he undertook it. [10] But the first draft of his work seemed to meet exactly the needs of atomic bureaucrats, and it was published without careful scholarly review that might have tempered overstatements and corrected miscalculations that eventually required retraction.

The first flush of publicity about the Rasmussen research tended not to go beyond the summary that prefaced nine volumes of data; much of the public probably heard no more than the single sentence that "a person has about as much chance of dying from an atomic reactor accident as of being struck by a meteor." That arresting comparison, based on mathematical models developed for the nation's space flights, seemed exaggerated to experts in probability and based on dubious assumptions to nuclear engineers. Two separate reviews by the Atomic Energy Commission's own staff suggested errors and omissions that ought to have muted the public relations blitz at publication. The scientific community's reception of the report was reserved and then, after study by several professional groups, quite critical, particularly of the summary that had attracted the media's attention. Finally, early in 1979, the NRC officially backed away from the document it had proudly hailed a few years before. WASH-1400, said the NRC's press release, underestimated the possibility of a serious accident, contained some numerical errors and miscalculations, had been inadequately reviewed before publication, and, particularly in the summary, was unreliable. "The Review Group concluded," said the NRC, that the summary had "lent itself to misuse in the discussion of reactor risks," which was surely a passive and undramatic way of disavowing the comparison to death by falling meteors. Fortunately, the NRC added, an internal study revealed that the agency itself had never relied on the report in the regulatory process anyhow; the commission's new stance would have no practical consequence either on operating reactors or on those proposed.[11]

Perhaps not. But the Rasmussen Report had been used extensively in the nuclear industry's effort to create a receptive climate for reactor development. Lawyers for utilities, as well as public relations employees, had cited WASH-1400 to serve their purposes. A federal judge in North Carolina wrote his own critique when he found some probability that a future nuclear accident might occur. He did not know what the odds were; "the court is not a bookie." But for Judge James McMillan, the question was not "whether a nuclear catastrophe" was "more or less likely than a tornado, or earthquake, or collision with a comet." Rather, he said, "the significant conclusion" was "that under . . . odds quoted by either side, a nuclear catastrophe is a real, not fanciful possibility."[12]

The NRC disavowed the Rasmussen Report about two months before the accident at Three Mile Island in March 1979 heightened public concern about nuclear safety. Later, the NRC and the nuclear industry would boast that the absence of a devastating explosion and traumatic injury at Three Mile Island proved that existing hardware and regulations reliably protected the public. At the time, however, members of the regulatory staff provided contradictory advice and confused interpretations and betrayed an incomplete understanding of events. Although there were no emergency plans, the governor of Pennsylvania and federal officials suggested that a temporary evacuation of some residents of Middletown and nearby communities might be prudent.

In the aftermath of the accident, following the recommendation of the President's Commission on Three Mile Island, both Congress and the NRC decided that future evacuations ought to be planned. The agency's new regulation, as Chairman John Ahearne summarized it for Congress, extended mandatory planning beyond the LPZ to "an area 10 miles in radius for exposure to a radioactive plume . . . and . . . about 50 miles in radius for food that might become contaminated." The commission understood that the failure of state and local governments to prepare plans might impede or even prevent the licensing of some reactors. But the NRC noted that various other prerogatives of local government, from zoning to rate regulation, already implied a capacity to inhibit operation, so the new regulation relinquished no authority the federal government could have retained.[13]

In effect, the new rule overturned the Seabrook appeal board's decision that plans beyond the LPZ were not required. Counsel for the states of New Hampshire and Massachusetts and for the Seacoast Anti-Pollution League (SAPL) and the New England Coalition on Nuclear Pollution (NECNP) quickly recognized the advantage the change conferred on the plant's opponents. Many questions about the Seabrook site, evaded or presumably settled earlier, were open again, and intervenors promptly reasked them. These same intervenors were later to protest vehemently changes in rules governing emergency planning. But this initial modification gave them legal leverage that was enormously advantageous. Changed rules did not invariably work to the advantage of utilities, and this one—in the middle of PSNH's effort to build Seabrook Station—was a substantial, even crippling, disadvantage.

A change in the regulation betokened no change in attitude, wrote an SAPL observer after attending a workshop on emergency planning with representatives of utilities and federal agencies. Although the Federal Emergency Management Agency (FEMA) approached the problem substantively and seriously, the utilities treated it as "an exercise in public relations." Planning, for the corporations, consisted "mainly of . . . fancy flow charts, diagrams, job titles," and slide shows. Plans were prepared, but "the mindset that serious nuclear accidents are extremely unlikely is as pervasive as ever." Several utility officials "expressed a . . . veiled resentment" at the expenditure of time and money "on planning for what they considered to be the extremely remote possibility of a serious accident." (Thomas Dignan once remarked that a safe evacuation after Three Mile Island "could have been accomplished on foot by a burro path.") The FEMA staff, the SAPL observer noted, had a much more constructive approach. [14]

Local officials in communities around Seabrook Station also began to consider workable emergency plans. Civil defense directors in Kensington and Hampton Falls, selectmen in Seabrook and Salisbury, the planning board of Portsmouth, and state officials from New Hampshire and Massachusetts sent inquiries or resolutions off to Washington. Much of this correspondence focused on the time required to evacuate the seacoast and on the estimated number of cars and people on which those timetables rested. Without conspicuous success, PSNH tried to emphasize other resources for the public's protection, such as the plant's design and construction. Public officials and intervenors were neither persuaded nor diverted from their concentration on emergency plans. [15]

The Seacoast Anti-Pollution League and the Commonwealth of Massachusetts formally asked in 1980 that construction be suspended until emergency plans had been submitted and approved. Jo Ann Shotwell, who prepared the case for Massachusetts, argued that safe and timely evacuation was impossible under any circumstances and clearly so in the summer. Since the new rule made eventual operation dependent upon plans she contended could not be drawn, she believed prudence required litigation of the issue before more money was spent. To permit construction to continue might cost owners millions in potentially unproductive investment that would eventually tempt the NRC to ratify a fait accompli.

Two commissioners thought Shotwell's case well made. "Seabrook poses difficult, and perhaps unique, emergency planning problems," wrote Victor Gilinsky in September 1981.

> We should begin to seek solutions now, not some years from now, when the plant is almost ready to operate. Moreover, at that time it will be much more difficult . . . to require remedial measures which could delay plant operation.

The "remedial measures" Gilinsky had in mind included major improvements to the region's highways; without such modifications, "Seabrook's operation may be contingent on restricted use of the beaches."

Gilinsky's predictions were more accurate than many that litter the record on emergency planning for Seabrook, but they were made in dissent. His colleagues on the commission, except for Peter Bradford, declined to review the decision of Harold Denton, director of licensing, who permitted construction to continue and who was not concerned about potentially wasted money: A loss of investment was "the risk every holder of a construction permit carries."[16]

Neither PSNH nor the NRC reacted to the shifting opposition that gradually developed after Three Mile Island. Accustomed to dealing with individual environmentalists and civil disobedients, proponents of Seabrook Station were slow to recognize that skeptics who held municipal or state office had local expertise (to say nothing of local constituencies) that deserved at least respectful attention. To achieve public credibility, after all, emergency plans would ultimately need the support of police chiefs, selectmen, civil defense directors, school principals, and others who shaped local opinion. Those were precisely the people who were asking questions when emergency planning surfaced as a major issue after Three Mile Island. And, in local matters, they did not defer either to the wisdom or to the authority of a utility executive or a federal bureaucrat. Their disposition to cooperate was not enhanced when what they believed to be legitimate concerns were ignored or dismissed. Perhaps the eventual impasse on the issue was always inevitable, but proponents of Seabrook Station did little in the early stages to avoid it. As one police chief remarked in town meeting when a resident asked why he had not participated in the design of emergency plans for the town, "Those people don't listen."[17]

Nor did they always reply. In October 1981 Governor Hugh Gallen of New Hampshire wrote Nunzio Palladino, then chairman of the NRC, seeking guidance as the state began preparation of emergency plans without which Seabrook Station could not be licensed. Gallen posed several specific questions and asked for standards against which plans should be measured in order to satisfy the NRC's requirement that adequate measures "can and will be taken" to assure public protection. How much time did authorities have, Gallen asked, to complete evacuation, and what sort of shelters should be available for those on the beaches? Could the NRC set levels of radiation exposure that must not be exceeded to accomplish "adequate protection"? What percentage of people in the region must receive prompt notification to satisfy that requirement?

More than six months later, Palladino sent seven pages of nonspecific answers to Gallen's queries. The net of Palladino's letter was that emergency planning was the state's problem, and the NRC would simply assess the result. His answers referred to federal regulations—the imprecision of which had been the basis for Gallen's questions in the first place—but did not provide the prescriptive guidance the governor had sought. "The particular sheltering options available," for instance, Palladino said, could "best be addressed by the licensee and the local governments." Or, there was no "minimum percentage of the population that must be notified" in order to comply with the agency's standard of "prompt notification." When the NRC completed its "in-depth analysis of the licensee's evacuation time estimates," Gallen would get a copy. The letter was neither prompt nor helpful.[18] And, in light of Palladino's remarks at the ACRS meeting in 1974, it was more than a little ironic.

Members of the NRC staff, meeting in the spring of 1982 with people from Yankee Atomic, were no more knowledgeable than Chairman Palladino. At the outset, indeed, the staff seemed not to know what the emergency planning regulation called for or what anyone must do to meet it. A cooperative spirit prevailed, and future helpful hints were promised, but what was then available was meager. James MacDonald asked what had to be finished before the owners could apply for a low-power license. "The policy on that . . . is not, I think, cast in concrete." MacDonald tried again: What must be done to inform transients? Perhaps, he was told, signs might be posted on the beaches, "saying, in case of a nuclear accident, do this or that, or call here." Maybe motel owners would help. Eventually someone suggested that the Federal Emergency Management Agency, which would have to pass on the plan, might have an opinion. What about the required annual exercise or drill, MacDonald asked, "a big-ticket item" that would involve participation by state officials and others over whom the company had no control? How could he promise their participation? Well, one of NRC's lawyers said, "you can . . . make a statement saying that you have provisions as long as others agree." So, MacDonald paraphrased, "We . . . make the commitment subject to agreement," which, he added, was "not much of a commitment." About the only specific advice MacDonald got was that it was "harder to fix emergency plans and procedures that started in the wrong direction than to get them right from the beginning." There must have been some annoyance in MacDonald's terse summary: "Do it right the first time."[19]

But, of course, it could not be done right the first time. New Hampshire engaged consultants to prepare prototype plans. These documents satisfied almost no one except PSNH, which welcomed anything that promised to reduce licensing delay, and the state's civil defense hierarchy, which had to sponsor them to FEMA. Several New Hampshire legislators unsuccessfully went to court to block the state's contract with the consultants, and local officials, both in New Hampshire and in Massachusetts, explicitly disavowed the firm's work.

The document for Newburyport, a city of about 15,000 south of Seabrook in Massachusetts, became a trial balloon. Hoping this plan might serve as a prototype, the authors submitted it to the NRC and to FEMA for comment and stirred up more than they bargained for. It was hardly tactful to send copies to local and state officials some weeks after federal functionaries had received the material, and ruffled sensibility may have contributed to the sharply critical response. Civil defense authorities in Massachusetts implied that future federal-state cooperation might be in jeopardy if the NRC accepted the plan, and Assistant Attorney General Shotwell set to work on a detailed critique.

Her case rested on more than local pride and irritation. The Newburyport plan required police officers to establish traffic-controlling barricades and provided for warning the public with sirens and radio messages. People could stay indoors, protection planners called "sheltering," or evacuate by driving their own automobiles south to Peabody. The plan had no provision for those unable to transport themselves and lacked directions to Peabody's reception and decontamination centers, which were not identified. It was, at best, an outline, which SAPL said "failed to satisfy a single one of the regulations." To be sure the shortcomings were thoroughly aired, Shotwell sent the NRC twenty-five pages of legal objections, all of which were unnecessary because FEMA decided that the model would not suffice and that a specific plan for the city would have to be forthcoming. [20]

The squabble over the Newburyport plan identified some of the issues and tested arguments for the full-dress hearings held late in the summer of 1983. With the aid of a growing number of volunteers familiar with regional peculiarities, intervenors scrutinized plans that admittedly had little local variation. Traffic control, for example, appeared to depend upon police forces many towns did not have. Part-time or volunteer town officers or firefighters could not be assumed to be available. State employees, including police, would not be able to substitute, as the plans promised, because they already had assignments elsewhere. Hospitals designated as emergency treatment centers reported that they had no facilities and no special competence to treat radiation cases. People to operate reception and decontamination centers were not identified, let alone trained. Towns had no means of measuring radiological exposure and no potassium iodide to dispense if public health officials authorized its use to block absorption of radioactive iodine in the thyroid. Hierarchical organization, which the plans detailed, depended upon prompt receipt of information, which fragile communication links made uncertain.

And then there were transportation and traffic, factors that from the start had made the Seabrook site problematic. Counting cars and parking spaces and motel rooms became a small industry in seacoast New Hampshire for several consecutive summers as consultants engaged by utilities, regulatory bodies, and intervenors collected evidence to sustain contentions. Computers processed data from aerial photographs and traffic surveys and disgorged estimates of the time required to evacuate beaches at various hours of day or night and under various weather conditions. Skeptics doubted that computer models gave adequate weight to accidents, cars without fuel, and the panic of parents who would drive to schools to retrieve their children regardless of what a plan directed. In any case, depending upon assumptions and sponsorship, the evacuation computations ranged from a minimum of six or seven hours to twelve or fourteen. When presumably expert opinions differed so widely, the question almost inevitably became one of whose expert was better, and the answer often depended less on credentials than on presuppositions. Local police and others with field experience tended to support larger estimates of population and longer estimated evacuation times.

Traffic, of course, was only part of the problem; some people would have no means of becoming part of the vehicular flow. How, for instance, would French-speaking Canadian tourists learn of the emergency and the prescribed routes leaving the area? School buses were the only quasi-public means of transportation in the region, and most of them were privately owned and operated by part-time employees. Neither drivers nor vehicles could be assumed to be ready at unscheduled times, and thus any evacuation of carless, disabled, or nondriving people, including those in schools and medical institutions, would not necessarily be prompt. Sheltering in place—that is, staying indoors— was the expedient the plans specified for such people until vehicles were available and evacuation was arranged.[21]

These practical concerns received legal formulation as contentions, which were litigated before the Atomic Safety and Licensing Board in August 1983. Thomas Dignan consistently maintained that emergency planning was not the owners' task but the state's, though they stood ready to provide assistance to either state if requested. When he commented on intervenors' contentions, Dignan ordinarily confined his remarks to technical, legal points that would narrow or eliminate the topic and thereby speed consideration. Later, in 1987, Dignan's clients would prepare plans for a noncooperating Massachusetts, but in 1983 they apparently hoped state sponsorship would expedite regulatory approval.

Dignan's self-effacing strategy turned the hearings into a confrontation between intervenors and the panel of judges, especially the redoubtable chairman, Helen Hoyt. Her tone and her rulings seemed calculated to send an unmistakable signal to the plant's opponents that she would keep proceedings moving toward the operating license she seemed determined to approve. She began by routinely denying several requests by intervenors and opened a running contretemps with Guy Chichester, now an officially designated observer from Rye, that would end only when she excluded him from the hearings several days later. To be sure, Chichester was provocative; after several PSNH officers were empaneled and their professional qualifications put in the record, Chichester remarked that these were the same experts who had promised the plant would be finished years before for less than a billion dollars. Hoyt told him he was out of order (which he was) and then explained that the stipulation about which he had commented dealt only with the witnesses' qualifications: "It has nothing to do with credibility or anything else." She meant, no doubt, that no one had abdicated the right to test credibility through cross-examination, but that was not the interpretation implied by Chichester's laconic "I see."[22]

During Jo Ann Shotwell's questioning of Robert Merlino, a consultant who had studied evacuation timing, the witness became almost extraneous to the tension between Shotwell and Judge Hoyt. The issues also seemed unimportant—Was the witness responsive or discursive, and did Hoyt disclose Shotwell's line of inquiry?—but the sharp exchange culminated in Hoyt's demand that Shotwell apologize. She refused and was expelled. Robert Backus completed her cross-examination of Merlino. Two days later, after a grudging apology secured her readmission, Shotwell charged that Dignan and NRC counsel Roy Lessy had used hand signals to coach Merlino and Thomas Urbanik, another NRC consultant. Her accusation, Shotwell said, rested on the observations of several conscientious, responsible town representatives.

The charge infuriated Hoyt, who lectured observers about ethics and professional conduct, calling their behavior "juvenile" and their accusation "frivolous." (She had earlier chastised lawyers in the case about their behavior too.) She would not listen when Roberta Pevear, the observer from Hampton Falls, tried to correct her misidentification of Merlino, who had not received coaching, for James MacDonald, who she believed had. Hoyt asked the witness who was present whether he had been coached and accepted his denial; she ordered an affidavit to be secured from another. She had, she said, been in a position herself to observe those testifying and had seen no coaching. She was officially saddened that proceedings over which she had presided would be marred by unsubstantiated and unprofessional accusations. Finally, she was shocked that Shotwell, the counsel for the Commonwealth, "would lend herself and her position to that serious an accusation."[23] Overwhelmed by the courtroom pyrotechnics was the substance of the testimony, which revealed that PSNH and the NRC were not in agreement about the time required to evacuate the beach area and that neither agreed with consultants hired by intervenors or with the chief of the Hampton police, the public official most directly involved; Hoyt decided his testimony had no relevance anyhow and excluded most of it.

When public officials could not persuade Judge Hoyt and the ASLB that emergency plans were seriously flawed, ordinary citizens did not stand much of a chance. But they appeared dutifully, when the judges scheduled "limited public appearances," during which citizens could tell the board what they thought without legal intermediaries. One or two people who worked at Seabrook Station thought the plant had splendid equipment and well-trained employees; it ought to be licensed without emergency plans since evacuation would never be necessary. But others, including several volunteers whose research furnished the factual skeleton for the arguments of intervenors, noted omissions, mistakes, and the need for participation by the communities concerned.[24]

Some of those communities began to assert their own right to determine the adequacy of radiological emergency planning, a claim Massachusetts had already made in effect. Sharing the preparation of plans had been the NRC's concession to local involvement, but assessment and decision were reserved to federal officials lest utilities encounter conflicting local laws that would halt completely the already staggering industry. Nevertheless, selectmen in Hampton Falls said they did not accept plans drawn for the town and told New Hampshire's civil defense officials that they would be notified when the town decided how to deal with radiological preparedness. When PSNH talked of installing sirens to signify an emergency, selectmen wrote the company that it would be informed when they had decided how to alert the public; meanwhile, no sirens. Kensington and South Hampton, outraged when attorneys for PSNH argued that towns could not raise legal objections to their own plans, disclaimed the state's plan and restated their objections. Suppose those "absurd" plans had called for evacuation by private aircraft or sheltering in subway stations, hypothesized the town officials; could they then suggest that the arrangements did not adequately protect their fellow citizens?[25]

The irony—and, for many people, the frustration—was that all the effort had no result; for PSNH, facing serious financial trouble, the fact that all the expenditure bought nothing must have been especially bitter. The legal controversy devolved into an argument about whether the NRC should remove Hoyt from the licensing board after she refused to disqualify herself. (In 1985 she was "temporarily unable to serve" and was replaced by Sheldon Wolfe, who continued to chair the board when safety with respect to the plant itself was at issue; Hoyt returned to deal with off-site issues, including evacuation, until 1987.) But the substantive concerns about emergency planning, for the moment, vanished. Corporate survival in the cash flow crisis of 1984 preoccupied PSNH.

New Hampshire withdrew the first set of plans for extensive revision, a process that required about three years and new consultants. Although many local officials kept a wary eye on the state's effort to draft new plans, the process no longer demanded rapt attention. The calendar for Seabrook Station was years behind and that for emergency planning was no different.

Postponement, of course, was not resolution. As it was obliged to do, PSNH duly reported to investors that any operating license continued to depend upon acceptance of emergency plans, which the civil authorities of several communities and the attorney general of Massachusetts opposed. Governor Michael Dukakis of Massachusetts, the utility noted in 1984, had said that any acceptable state plan must have the endorsement of all affected towns. That circumstance made the utility "unable to predict whether . . . completion or acceptance of the plans" would be delayed or even prevented, thereby precluding receipt of an operating license. Though PSNH's prose had a tone of determination, which perhaps resulted from unwavering federal support for nuclear power, management must have known that Dukakis's position in effect gave one Massachusetts town a veto over the plant's operation. The NRC might override a local obstruction, but such a ruling would predictably wind up in court. That was neither a cheery nor an inexpensive prospect.[26]

Familiarity with their own locality gave opponents of Seabrook Station an assurance in litigating emergency planning they lacked when other measures for protecting the public were at issue. With the exception of construction workers who had encountered inept design or witnessed shoddy craftsmanship, few opponents of nuclear power had direct or personal acquaintance with the plant and technology they battled. Their knowledge of safety systems derived from published research and reporting, from governmental investigations, and from remarks of disaffected dropouts from the nuclear establishment. Secondhand charges lacked the intensity of personal experience and the emotion connected with family and property. Intervenors who dealt with specifications of pipes or wires, with the chemical properties of radioactive by-products, or with the design of control panels sometimes seemed to have looked up facts to buttress a case against nuclear power that rested on other, more personal grounds.

In addition, the technical complexity of safety devices conferred a tactical advantage on utilities comparable to that intervenors enjoyed when emergency planning was disputed. Plant owners and the regulatory authorities with whom they were often allied had ready access to professional nuclear engineers, whereas intervenors often lacked funds to hire experts capable of refuting nuclear proponents. The amateurs and academics upon whom intervenors relied had not ordinarily made careers as nuclear experts; consequently, their testimony tended to require extensive, often hasty, and invariably ill-paid preparation. Proponents of nuclear power hindered that preparation by denying access to construction sites and proprietary information. And the NRC had several bureaucratic devices to obscure or postpone questions about safety, even when those questions came from the agency's own staff. Lacking data and expertise and blocked from contesting governmental policy, counsel for intervenors could not easily make an affirmative case. So they fell back on cross-examination, which was not always informed and which sometimes seemed only an effort to discredit witnesses, uncover contradictions, and secure unwary admissions. The strategy was neither amiable nor efficient, and it did not often work.

Intervenors' reflexive hostility to nuclear power usually confronted an equally reflexive support, an attitude on the part of nuclear advocates that the technology was safe and that challenges were inappropriate when not frivolous. The President's Commission on Three Mile Island found this complacency at least an indirect danger to the public. *"Fundamental changes,"* read the commission's very first recommendation, *"will be necessary in the organization, procedures, and practices,—and above all in the attitudes"* of the regulatory agencies and the nuclear industry. Several years later, investigators of the accident at Chernobyl identified that same assurance as a central cause of the disaster. "This was a very good plant," a Soviet official commented. But the operators "got too confident" and disregarded hazards that had for them become routine.[27]

Opponents of nuclear power plants claimed that the lessons of Three Mile Island had never been mastered, to say nothing of those to be learned from Chernobyl. But the NRC routinely deflected many technical and engineering challenges because they applied to more than one plant. Such issues, called "generic," would be resolved by a binding regulation in the future and did not require delay or suspension of construction or operation; compliance with the eventual rule would suffice. Yet the president's commission and observers outside the agency have charged that important safety issues have been postponed beyond any reasonable limit; the device has become a way to obscure technological deficiencies and to permit business as usual. Such "generic issues" as fire protection, the integrity of cooling systems, and the ability of emergency equipment to function under conditions of extreme temperature and pressure, have remained unresolved for a decade or more.

The temptation to ignore or evade such matters is acute. A decision might require licensees to turn off reactors and spend heavily to replace power and rebuild portions of plants. Affected utilities, not unnaturally, will wonder why expenditure is necessary if the plant has run in a presumptively safe manner for some time, and why the design, if flawed, was originally approved. Opponents will point to revised rules as evidence of bureaucratic incompetence and a fundamental failure to protect the public. Short of a crisis, such as Three Mile Island or Chernobyl, retrofitting, even to achieve public safety, has no powerful constituency. Daily tasks beckon, research does not seem urgent, and the NRC has allowed some of the most perplexing generic issues to drift without resolution, thereby permitting completed plants to operate for years in spite of reservations about safety components.[28]

David Okrent, a longtime member of the Advisory Committee on Reactor Safeguards, has described persistent pressure from the NRC, and the AEC before reorganization, to suppress misgivings about plant design, or at least to withhold them from the public. In the mid-1960s, the committee threatened public criticism of the industry's failure to devise reliable protection against loss of coolant in the reactor, a condition for which "meltdown" and "China syndrome" became the popular phrases. The AEC kept the committee's concern off the record, and when the ACRS sent a formal protest the AEC urged that it be withdrawn lest it "lead to misunderstanding by the public."

That experience, Okrent says, was not unique, and the result was the absence of a record upon which the ACRS could base subsequent objections when research went undone and misgivings unaddressed. Further, approval of one plant with unacknowledged reservations led inevitably to approval of others with similar deficiencies, because the ACRS had no basis for distinguishing one from the rest. Utilities could not reasonably be required to submit new solutions for problems that had presumably been solved or to add hardware that competitors had not been forced to purchase. Nor did the ACRS want to imply error in its own previous acceptance of similar designs. Thus, the lack of a public record and the risk of institutional embarrassment combined to persuade the committee to resolve doubts within the regulatory framework, to allow licensing to continue, and to leave the public unenlightened.[29]

David Okrent himself badgered witnesses in April 1983, when the ACRS considered an operating license for Seabrook Station. In particular, he wanted to know, and could not find out, whether safety devices installed at Seabrook would survive a hydrogen burn. Other members of the committee pursued other questions, somewhat less aggressively than did Okrent, to similarly inconclusive results. One member of the group asked about the effect of rapid temperature change on pipes in the cooling system, a matter the committee decided was generic but urged the plant's owners to investigate in any case. The interrogation was usually gentle and the atmosphere apparently one of colleagues who shared a sense that they were going through motions toward an agreed conclusion, though the ACRS was clearly serious about the enterprise. The chairman, for instance, asked a witness about the pulse rate of an operator who pulled a switch to correct a serious reactor malfunction, "and the handle comes off in his hand," as had occurred during a Seabrook drill. When the laughter subsided, the witness replied that the handle had been replaced in seconds. He also assured the ACRS that an earlier decision to use pistol-grip switches was under active review, a reply that seemed to satisfy the committee. A majority of the committee, indeed, found the promise of future information a sufficient response to any concern. The ACRS filed no objection to a proposed operating license for Seabrook Station. [30]

Intervenors, of course, had less confidence in the commitments of Seabrook's owners. There was "absolutely no precedent," as counsel for the New England Coalition on Nuclear Pollution put it, "for accepting promise in lieu of performance." The specific argument was about whether management had sufficient resources to assure quality control, but the same suspicion applied to several other matters. PSNH told Dana Bisbee, an assistant attorney general of New Hampshire, that eventually a postaccident monitoring system, as required by the NRC, would be fully deployed but that licensing need not be delayed until the system was actually in place. That pledge, Bisbee retorted, proved his point, which was that the corporation had not met regulatory standards. He did not want to monitor the company's future performance; he wanted compliance before operation. But a pledge, the NRC held, sufficed for the moment, because a license could eventually be refused if Seabrook's owners did not live up to their obligations. [31]

In a sense, the NRC's entire system of enforcement relied on promises and trust. The agency did not supervise routine construction or the performance of contractors but relied instead on a station's owners to furnish oversight and quality control. Inspectors checked on management, specifications, and documents rather more than on electricians and welders. But in 1984, when Commissioner James Asselstine visited Seabrook and received a letter and photographs detailing serious construction deficiencies, the NRC's inspectors left their desks. Several days at the site revealed cracks and leaks in the concrete and improper storage of motorized valves, as the letter had charged. But the concrete was structurally sound, the leaks had been stopped more than a year previously, and the valves worked in spite of exposure to weather. The inspectors could not verify allegations of defective welding, bent pipes, and burned valves. The charges, in other words, were not substantiated, which was not quite a positive finding that the plant would protect the public's health and safety. [32]

Many of the same charges recurred two years later, together with some new ones, in a thirty-page critique of the NRC's investigatory effort. This manifesto rested on affidavits collected by the Employee's Legal Project (ELP), an organization that provided anonymity and legal support for Seabrook construction workers. They described unauthorized construction techniques, sloppy craftsmanship and design, defective quality control, and drug and alcohol abuse. Piping inspectors told of "cold pulling," a euphemism for bending or forcing pipe to make joints and a practice that might impair the strength of the pipe itself or of welds. Other workers described flaws in cement work, improper installation of valves, and debris that might block piping essential for firefighting. The ELP noted in 1987 that NRC inspectors had often simply accepted management's assurances without investigation and had "sidestepped" and "misconstrued" the ELP's allegations. The agency cited rules prohibiting cold pulling, for example, to prove the technique had never been used.

Further, the ELP continued, details of wrongdoing were only illustrative of more general malfeasance and not the organization's central point. Rather, the ELP believed that management had failed to insure that the plant was built according to standards that assured public health and safety, and that the NRC had failed to exercise regulatory oversight. The ELP asked that experts independent of the NRC and of Seabrook's management conduct a thorough and impartial investigation of the construction of Seabrook Station. No one dignified the request with a response, partly, perhaps, because independent audits of nuclear plants in Michigan and Ohio had led to their abandonment. Antinuclear activists have claimed that the Reagan administration, in an effort to bail out the beleaguered industry in 1984, ordered the NRC never again to authorize an independent investigation. [33]

Whistle-blowing workers at the nuclear plant at Shoreham on Long Island could have warned their New Hampshire counterparts of a frosty welcome from functionaries at the NRC. A conscientious Shoreham quality control employee told of corporate pressure to "get the plant on line" even if to do so required winking at deficiencies. NRC inspectors were not interested. "All plants have problems" was their attitude. "It's no big deal."[34]

The tendency of engineers at the NRC to prefer the word of other engineers to accounts of disgruntled employees was, in fact, "no big deal." The mutual confidence of people who made a career of nuclear power rested on shared convictions that research, facts, and experience were on their side. Occasional defects were essentially minor and susceptible to some relatively easy fix; in no way did they require a reexamination of the premise that nuclear power was both safe and essential. That outlook, combined with high bureaucratic status and extensive technical training, did not always result in patient or tactful responses to questions from tale bearing welders or uneasy housewives.

One of the high-tech tools that contributed to the confidence of the nuclear establishment was probabilistic risk assessment. Although the NRC in 1979 had backed away from the extravagant claims of the Rasmussen Report, the agency had not disavowed the methodology upon which that report rested. In 1984 PSNH submitted what the consultants who prepared it claimed was a state-of-the-art report using the most up-to-date statistical techniques. The document resulted from "a highly scientific endeavor requiring the highest levels of technical competence and integrity." Computers had digested "dozens of logic models" and made "tens of thousands of calculations" to evaluate "billions of accident sequences" and "millions of pieces of data" at a cost of "hundreds of thousands of dollars." The outcome of this elaborate and rigorous study, which had been reviewed by a distinguished advisory panel including Norman Rasmussen, was "an independent assessment of the risks associated with the operation of Seabrook Station."

Those risks were predictably trivial and would increase if the plant were to burn natural gas instead of relying on nuclear fuel. Since there was no chance of a break in the plant's containment, there was "no appreciable risk of early fatalities from the operation of Seabrook Station." Only an earthquake of a severity never observed in the region or an accident that somehow allowed coolant to bypass the containment building (a flaw that had been "designed out" of the system) would require modification of that reassuring conclusion. The report quantified the risk: Accidents leading to early deaths from cancer might in the worst case occur once in every 3,000 to 5,000 years, and more probably in the range of 7,000 to 20,000 years; accidents that would result in serious injury were even less likely. Since Seabrook Station would produce power for fewer than fifty years, the report concluded that operation would not add at all to hazards already inherent in residence in the region. [35]

Risk assessment, for all the trappings of computer runs and numerical precision, requires an ability to imagine what might go wrong and depends upon arguable assumptions, such as the one in the Seabrook study that coolant would not escape containment. Critics have noted that most accidents are, after all, unimagined and that the one at Three Mile Island was, by the NRC's definition, "incredible." Furthermore, as Charles Perrow points out in *Normal Accidents*, risk assessment is value-neutral. That is, the technique does not distinguish fifty lives and a hundred million dollars in property lost in a nuclear accident in one community from fifty deaths around the world and a million accidents each involving the destruction of $100 in property. Risk assessors, Perrow concludes, "have a narrow focus that all too frequently (but not always) . . . supports the activities" that society's "elites . . . think we should" undertake. Their studies focus on "dollars and bodies" and ignore "cultural and social criteria." The public, particularly the public living with a neighborhood nuclear plant, may have a rather different emphasis. [36]

Robert Rader, who prepared a paper on emergency planning for the Washington Legal Foundation, thought the public's concerns ought to be ignored to accelerate approval of nuclear licenses. The NRC's hearings on off-site safety and emergency planning should simply cease, and plans approved by the Federal Emergency Management Agency should be treated as presumptively valid; FEMA had the expertise and the responsibility to shape emergency plans, and the NRC's rules provided an unnecessary forum for antinuclear agitators. An executive order, Rader claimed, could reduce the radius of the planning area from ten miles to two and improve the entire process without legislation.

The point was not just to rearrange the tasks of federal agencies but to minimize and perhaps eliminate public participation and the influence of state and local governments. The NRC's hearings, Rader said, were unsuccessful "even as a public relations measure" and tended to diminish "public understanding of nuclear power and public confidence in plant safety." Furthermore, local governments frequently conditioned their approval of plans on a utility's willingness to furnish equipment and supplies that taxpayers ought to have purchased; hearings enhanced the "utility's vulnerability" to this official blackmail. The process ought to be streamlined, efficient, and rest on the competence of experts rather than the clamor of ill-informed citizens and their timid representatives. [37]

Rader's paper did not cause the utilities to urge changed rules on the Reagan administration. But sometime after January 1984, when nuclear power producers endured a series of setbacks, a task force in the executive branch of the federal government set out to prop up the reeling industry. Investigative reporters have not uncovered the entire agenda of this group, though efforts to assist PSNH and Long Island Lighting were part of the acknowledged purpose. [38] At about the same time, beset by financial crisis, PSNH temporarily halted work at Seabrook and shuffled its management. One federal agency, the Rural Electrification Administration, through the New Hampshire Electric Cooperative, helped PSNH secure additional capital that eventually enabled completion of one reactor at Seabrook Station, though there is no direct evidence linking that action to the federal task force. The existence of the group, in any case, sent a clear signal to regulatory authorities that the Reagan administration supported the industry. One way to acknowledge that support would be to ease the rules on emergency planning, as Robert Rader had suggested.

But those rules had a political constituency that the Reagan administration itself placated during the campaign of 1984 in order to help a struggling Republican congressman. Secretary of Energy Donald Hodel assured Congressman William Carney that the federal government would not impose its authority on his Long Island district "in matters regarding the adequacy of an emergency evacuation plan" that state and local governments opposed. The president expressly and publicly echoed Hodel's pledge. [39] Since authorities in Suffolk County and New York State had frequently declared that no plan could assure safe evacuation of the region around Shoreham, Reagan's statement appeared to close that controversy, though of course closure turned out to be temporary. And since several Massachusetts communities within ten miles of Seabrook, with the backing of the governor, had taken a position like that of Long Island officials, Seabrook too appeared unlikely ever to operate.

But the licensing process went ahead, as if somehow divorced from presidential promises and policies of the Commonwealth of Massachusetts. With the firm support of Governor John Sununu, New Hampshire civil defense officials refined in 1984 and 1985 Seabrook emergency plans that had been in abeyance since 1983. Local officials in several of the affected communities renewed their reservations about those revisions, however, and forced the state to devise and then impose a scheme that would be effective, its authors claimed, even without the cooperation of local officials. Selectmen in Hampton, for instance, reviewed in 1985 a draft of the plan proposed for their town, after which three members of the five-person board protested that the document could not be described as a serious attempt to protect "an endangered citizenry." The town lacked equipment, personnel, facilities, and time to fulfill the responsibilities assigned. The scheme bore almost no resemblance to local reality and seemed to have been written simply to fill a licensing requirement, not to protect the public. The state had not improved the road network on which prompt evacuation depended. Finally, the plan rested on the performance of the plant's owners, for whom citizens of Hampton had a "general distrust" because of a multitude of unkept promises. Richard Strome, the state official responsible for emergency planning, waited a month before sending a soft and reassuring response.

Strome had to answer a good deal of irritated mail. Officials in Hampton Falls were annoyed that he had sent a plan for their town to FEMA without allowing them a chance to review it. Selectmen in Hampton and Rye sent the same complaint directly to FEMA. Authorities in Rye were particularly incensed that they were obliged to submit legal objections to the plan in little more than a month, which was about half the period federal authorities had had for their consideration.[40]

Although New Hampshire had revised the plans since the first edition in 1983, they provoked the same objections, which boiled down to an assertion that they would not work. The themes recurred in document after document that piled up in the files of the licensing board: Estimates of population and traffic, especially at the beaches, were too small; consequently, calculations of the time required for evacuation and predictions of the radiation to which people would be exposed were too low. Provisions to shelter schoolchildren, the infirm, and those without access to transportation were inadequate; alarms would not alert the deaf or those who closed their windows or had noisy television sets, air conditioners, or furnaces. Inadequate highways would plug with ice or with buses bucking evacuation traffic, seeking those who could not drive themselves. Private automobiles would run out of gasoline or into one another. Towns lacked people to tow disabled vehicles, control traffic, drive buses, plow snow, and staff decontamination centers, and there were no radios, vehicles, and other equipment to perform tasks the state had assigned. The state's promise to supply buses, ambulances, police, and other personnel was unreliable, since both people and equipment had assignments elsewhere. Once those general shortcomings were remedied, then specifically local matters—like the inadequate shelter of glass walls at the school or the configuration of a particular intersection or the reliability of arrangements with a towing company that had gone out of business—could be addressed.[41]

Couched in legal prose, Thomas Dignan's reply to these objections was essentially "So what?" Intervenors were, he said, trying to relitigate the siting decision they had lost more than a decade before. Emergency planning regulations did not "impose new performance or siting criteria," nor did those regulations "require . . . a demonstration of . . . perfect safety." Neither New Hampshire nor Dignan's client had to alter facilities at the site or build hospitals or roads or add to local police forces or fire departments before Seabrook Station could legally generate electricity. Regulations required only a good-faith effort to allocate existing resources in advance of a possible emergency, not a flawless scheme that absolutely protected everyone who happened to be within ten miles of Seabrook Station. If a nearby town, such as Hampton Falls, did not employ a full-time police officer, and consequently felt insecure, then Hampton Falls should hire additional police; if Kensington needed a supply of radiation-blocking potassium iodide, then the town should buy some; if South Hampton lacked equipment officials believed necessary to evacuate the population, then they should purchase it. If local governments disagreed with the state government about the feasibility of arrangements or the condition of highways, those arguments did not involve the owners of Seabrook Station and ought not to keep them from using it.[42]

A few months later, two members of the appeal board explored Dignan's intriguing interpretation of the regulation on emergency planning. Suppose, Gary Edles mused, that "the facts turn out that you can't shelter" all the people at the beach on a summer day and "you can't evacuate them." "Are you telling me," Edles inquired, "that's show biz?"

"That's show biz," Dignan replied.

"And you want me to sign my name to that?"

"Yes," said Thomas Dignan. His clients were obliged to do the best they could with resources at hand. If "no action or very little action is possible given the existing state of affairs, that is that."

"Wait a minute," Alan Rosenthal, the presiding judge, interjected; "if we adopted" that position, "I could never turn up in the Common-wealth of Massachusetts to visit my sister again." "Now are you telling me," Rosenthal continued,

> that if it turns out that there is absolutely nothing that can be done for these people on the beach . . . to ensure that they do not get extensive radiation exposure, that is just too bad [?] Just as long as your plan does the best it can do, and if the best it can do leaves these people in a position where, given a certain accident, they are going to receive considerable radiation exposure, that, as Mr. Edles quaintly put it, is show biz.

Dignan evaded a direct answer and suggested that previous proceedings had accepted the Seabrook site; that decision could not be retried in the guise of emergency planning. But Rosenthal was not buying that argument either. He had, he said, been in the case from the outset, and emergency planning for the beach had never been addressed. Dignan said the plant's design and construction protected the public, including bathers at the beach. That was step one, Rosenthal agreed, but emergency plans were required as well.[43]

The way those plans appeared to pit local governments against the state puzzled Emmeth Luebke, one of the members of the Atomic Safety and Licensing Board that met in March 1986. He asked Ben Lovell, Kensington's civil defense director, if the town had attempted to alter the state's plan. Yes, Lovell said, but the town had given up when the state adhered to its own point of view whenever there was disagreement. If you do not participate, Luebke continued, how can you object? Lovell answered that litigation seemed the only effective way to get the state's attention. Luebke suggested to Dana Bisbee, who represented New Hampshire, that he consider and, where possible, incorporate local objections to improve any forthcoming revision of the state's plans. Lovell remained skeptical; some of the objections being discussed in 1986 had been copied from ones he had filed in 1983, and they had not had much impact on the process. Attorneys for Rye and Hampton noted that their clients had also found New Hampshire inflexible.

Luebke came to the point: Could not, he asked, the various parties get together and resolve their differences without using the licensing board? Town representatives, in response, displayed rigidity similar to that they decried in the state. Attorneys for Rye and Hampton took the position that no plan, by whomever devised, could adequately provide for the health and safety of their communities; that left little room to negotiate. If, said J. P. Nadeau of Rye, the ASLB had a proposal, the town would consider it. Helen Hoyt, the chief judge of the panel, said the board was offering neither to draft new proposals nor to negotiate among the litigants. Rather, she simply hoped to reduce the number of disputes and save everyone time and money. But, she said sternly, the board had to have a plan. Paul McEachern, attorney for Hampton and soon to be the Democratic candidate against Governor Sununu, could not let that remark pass, since it implied that some plan would be accepted. You could, he reminded Judge Hoyt, find that none of the proposed schemes adequately provided for the public's health and safety. That was so, she agreed, but she lacked the authority to overturn a siting decision already made and a construction permit already issued, a remark that responded to requests no one had made, echoed an argument of Dignan's, and seemed to foreshadow the rubberstamp judgment McEachern feared.[44]

While lawyers wrangled in 1986, the Federal Emergency Management Agency gave New Hampshire's plan a practical test. Massachusetts, of course, declined to participate in the drill, as did municipal officials and employees from seven New Hampshire communities, including Hampton. The February date not only ruled out beach traffic but schoolchildren as well, because schools were on vacation. The routine of hospitals in the area was not interrupted, so conditions did not exactly replicate those that might obtain in an emergency. Nevertheless, FEMA deployed four dozen observers to watch the response to a simulated accident.

The report of deficiencies exceeded 200 pages. The system for alerting the public had not been tested, and several prescribed communications were never delivered; an hour after the order to evacuate, for example, the public still lacked instructions about where to go. That omission was probably fortunate, because if 20 percent of those who were supposed to go to Manchester had appeared the facility would have been overwhelmed; perhaps that was the reason directional signs were few and inaccurate. Evacuation orders were not coordinated with instructions about bus routes, which made little practical difference because there were too few buses and even fewer drivers. Drivers who did participate had no maps and inadequate directions; some could not find their destinations, and others had insufficient fuel; none had training in radiation detection or protection. Plans to use state personnel to replace people from nonparticipating communities broke down completely when the state had no substitutes. State police, for instance, were supposed to cover forty-four traffic points, but only thirty-five troopers were assigned to the region, which did not leave many to fill in for absentees from Hampton or Hampton Beach or Rye or other towns that refused to play what they called a charade. The Coast Guard failed to alert ships at sea within the prescribed time. The supply of ambulances was inadequate, even if those that were supposed to travel a couple of hours had bothered to make the trip. Officials could not be sure their surveys had identified everyone in the area who needed ambulances or other special transportation anyway. And so on for pages. FEMA worded its conclusion more tactfully, but the message was that the plan needed a wholesale revision that would delay Helen Hoyt's scheduled hearings for some months.[45]

Dignan fired off a legalistic protest holding that FEMA was measuring Seabrook's plan against too rigorous a standard. To the ASLB, Dignan argued that the flaws FEMA identified were only defects in the exercise, not in the plan itself. The shortcomings were in no way fundamental, he said, and the board should simply assume they would be fixed, a posture that was too complacent even for the NRC staff, which noted that plans could hardly be found to be adequate if they could not be exercised.[46]

Thomas Dignan evidently believed emergency planning dictated a different legal posture from that he had adopted in earlier phases of the prolonged litigation. Then he had been the master of detail and argued the substance of issues as well as the law: The cooling system not only met the letter of environmental law, for example, but positively demonstrated his clients' sensitivity to the ecological and recreational concerns of intervenors. By contrast, when he dealt with emergency planning, he brushed aside concerns about traffic and buses and too few police and relied instead on a narrow interpretation of law and restrictive definitions of his clients' obligations. However correct his reading of the law, Dignan's strategy seemed unresponsive to the concerns of an articulate segment of the public and to the public officials who were attentive to it. One of his legal maneuvers, for instance, resulted in an order excluding nine towns in New Hampshire and one in Massachusetts from participation in the still unsettled battle over emergency planning. The order provoked the reproving intervention of political leaders in both states, some of whom supported a license for Seabrook; the order was eventually modified.[47]

In part, Dignan was no doubt reacting to mounting financial pressure on Seabrook's owners, which reduced their tolerance for endless litigation. In part also, Dignan's hands were tied because PSNH and the other utilities did not write the plan in question. They provided support, consultants, and advice, but the state had to sponsor and defend the result from the detailed critique of informed opponents. In one sense, Dignan was not even a party to the dispute, though of course the plant would remain idle until it was resolved. He tried to keep the calendar moving by applying in June 1986 for a license to test the plant at 5 percent of capacity.

Intervenors saw no need for haste. Both the Commonwealth of Massachusetts and the NECNP asserted that the NRC's regulations required approved emergency plans before a low-power license could issue, a reading the staff disputed on the basis of an earlier ruling permitting low-power testing at Shoreham. But plans for Seabrook would be forever incomplete, argued Carol Sneider for Massachusetts, because her state had decided to submit none. Without those plans, there could be no operating license. Except as a step toward that license, low-power testing had no significance. The step need not, and should not, be taken.

Jerry Harbour, a member of the ASLB, understood Sneider's point and was outraged. As he saw the situation, Massachusetts was intentionally obstructing accepted legal procedure and then asking him to approve. "This is an affront to the judicial process," Harbour charged. Further, he pointed out, the Commonwealth's officials had indicated that, if Seabrook were to operate, they would not even prepare, let alone carry out, emergency plans to protect the citizens. That statement, Harbour said, was "appalling, and in direct contradiction" to the concern for the health and safety of the public the state professed.[48]

Jerry Harbour may not have read Robert Rader's piece advocating changed rules for emergency planning. But the judge's expostulation certainly provided an on-the-record manifestation of the NRC's frustration with existing procedure, which inadvertently permitted states and municipalities to block national energy policy. Those charged with carrying out that policy thought the anxieties of the public overblown, even if sincere, and the actions of politicians opportunistic and ill informed. Yet neither the NRC's authoritative information nor its regulatory muscle could overcome popular misgivings and political impediments. The agency, the utilities, the nuclear industry, and its advocates looked for pretexts for changing the rules or for making exceptions to them.

As part of their effort to evade the barrier erected by Massachusetts, Seabrook's owners commissioned a new risk assessment study. They expected this research to demonstrate that the plant was so safe that emergency plans beyond a two-mile radius were unnecessary, a distance that would make Massachusetts irrelevant. Although the study was not publicly acknowledged, opponents knew of its preparation, and Congressman Ed Markey, who chaired a subcommittee on energy and whose Massachusetts district included communities within the ten-mile radius, warned the chairman of the NRC not to underestimate the state's reservations or to attempt to circumvent them "by regulatory artifice." Markey believed the NRC had encouraged New Hampshire Yankee (NHY) to find a legal path around emergency planning regulations and had sponsored research with public funds to assist. So he was not surprised when the utility, in December 1986, filed a two-volume petition asking for an exception to the ten-mile rule based on precisely those "enhanced safety features" and "special circumstances" the NRC staff had suggested be emphasized. Those unique characteristics, the petition said, would achieve more protection against accidentally released radiation at one mile than regulations specified at ten. As written, the ten-mile requirement served "no health and safety purpose" at Seabrook but only blocked operation, a result clearly contrary to the intent of the rule. "The extra nine miles" only made "planning more diffuse, more costly, subject to more political vagaries, and less focused." Any "additional risk avoidance" to be accomplished by insisting on a full ten-mile zone was "so extremely low as to be properly characterized as negligible."[49]

Brisk as usual, Helen Hoyt ordered intervenors to file responses to the petition within a month, a requirement even the NRC staff thought unreasonable. The National Laboratory at Brookhaven, the NECNP said, had spent nearly a quarter of a million dollars and four months without finishing an evaluation; under those circumstances, as Robert Backus put it for SAPL, requiring a response within a month was "arbitrary and capricious, and inconsistent with the Board's duty to adjudicate matters . . . in an evenhanded way." Judge Hoyt was not moved. Her task, she said, was to determine whether the petition warranted the commission's review, a relatively simple assessment that ought not to require prolonged consideration.

Except for New Hampshire, which told Hoyt that Yankee had contracted to supply the state with plans for a ten-mile zone whatever the fate of the petition, intervenors filed predictable objections. The burden of their argument was that distance provided protection, a fact the NRC itself had acknowledged when adopting the existing rule. A one-mile zone, SAPL noted, failed to provide for the security guard at the plant's gate, which stood more than a mile from the reactor; submission of the request, Hampton charged, constituted an admission that Yankee could not comply with existing rules. An assistant attorney general of Maine called the petition an indirect attack on the regulation. [50]

The frontal attack came from the NRC itself. Early in February 1987, Congressman Markey released the text of a proposed rule, drafted by the NRC staff, that would allow nuclear plants to operate without emergency plans approved by state and local authorities. The issue, wrote William Parler, the NRC's general counsel, and Victor Stello, Jr., the executive director of operations, was one of "equity and fairness," when a "utility has substantially completed construction" and then encountered the "non-cooperation" of state and local governments. The "forced abandonment of a completed nuclear plant," they argued, would have "serious financial consequences" for "the utility, ratepayers, and taxpayers." Although wasted money alone could not justify any revision because courts had held that the health and safety of the public had no price, Stello and Parler claimed that their "more flexible" approach to emergency planning would give the public the same protection that the "relatively inflexible" existing rule provided. When that rule was adopted after Three Mile Island, they wrote, the commission expected the full cooperation of state and local authorities, which "in a few cases" had not occurred. The proposed change simply restored to the NRC the control over atomic energy Congress had intended to lodge there.

No legislation was required; the commission could amend its own regulations to permit "a full-power operating license . . . notwithstanding non-compliance" with emergency planning requirements if that noncompliance derived from the refusal of state or local governments to assist. To qualify, a utility had to demonstrate "a good faith and sustained effort to obtain . . . cooperation" and "effective measures to compensate" for the absence of governmental participation. Stello and Parler doubted that a drill, like the one conducted at Seabrook in February, would prove much, and they suggested that requirement be dropped.

Approval of a utility's plans by the NRC, Parler and Stello argued, would conclusively establish the feasibility of protecting the public.

Once protection was assured, conscientious public officials would no doubt abandon their opposition, do their duty, and help make plans even more effective. State and local leaders, in other words, would just adopt the judgment of a federal agency and subordinate their own convictions and interpretations of their official duties. Stello and Parler did not mention the police power, a legal thicket in which a court might discover that a state's obligation to protect citizens was equal to that of the federal government.[51]

Stello's prediction that this proposal would prove "controversial" was, as he knew, an understatement. Congressman Markey characterized it as both unconstitutional and "outrageous" and introduced legislation— which ultimately failed to pass—that would have explicitly allowed governors to veto licenses for nuclear plants. In the required public hearings on the rule change, one public official after another recorded opposition, including the two governors the amendment had been designed to circumvent. Mario Cuomo of New York was almost contemptuous of the scheme he labeled "absurd," "contradictory," and "palpably without rationale" except that of protecting the foolish investment of utilities. The commission had conceded in 1980, Cuomo reminded his audience, when the regulation on planning had first been adopted, that "the operation of some reactors" might "be affected . . . through the inaction of State and local governments." He cited President Reagan's 1984 promise that the federal government would not override regional officials and asked if the revision was "not a direct contradiction of the President's position." "Will you say," he continued sarcastically, that the president "didn't mean it? Or that he changed his mind? And if he did, when did he change his mind?" Preempting the states, Cuomo warned, would "squander" what remained of the commission's "limited credibility," undermine public confidence, and contribute to "the collapse of nuclear power."

Governor Dukakis was more restrained than Cuomo but not less adamant. Massachusetts had maintained for more than a decade, he said, that "the area around Seabrook . . . could not be evacuated in the event of a serious nuclear accident"; this was no surprise he had recently sprung on the Public Service Company of New Hampshire. The company's investment was the pretext for the change, which the governor called "the nuclear equivalent of cutting the number of lifeboats for the 'unsinkable' Titanic" because they would "make the voyage unprofitable." That, the governor admonished, is "not what this Commission was set up to do."

That's not what governors are elected to do. And that's not what Congress intended when it gave you—and me—the responsibility to protect the public health and safety of the American people.[52]

Unmoved either by sarcasm or by metaphors about sinking ships, the NRC voted to publish the proposal in the *Federal Register* and to ask for additional comments. One of the most important, and most reserved, of those comments came from FEMA, the agency charged with evaluating emergency plans. The issue, as FEMA saw it, was one of public policy that might have to be resolved in Congress or the courts; in any case, the proposal was not just a minor correction. Nor was FEMA ready to adopt the position of the authors that there would be no reduction in public protection. The lack of a required drill or exercise was a particularly important omission that would force FEMA to rely on conjecture in assessing proposed plans; the agency wanted a more reliable basis for judgment.[53]

Amending the rule on emergency planning might eventually enable Seabrook Station to operate, but officials at New Hampshire Yankee feared the process would take too long. They preferred, they told NRC staff members in March 1987, a quick approval of their petition to reduce the zone. They had not made the request frivolously, they said, or simply to eliminate the "vague and unsubstantiated" concerns of Governor Dukakis, which they had conscientiously tried to calm. Rather, the petition was "extremely well documented," rested on years of research, and warranted the approval that would advance the licensing process.

There was, Victor Stello suggested, yet another strategy. The NRC had agreed in the Shoreham case to consider utility-sponsored emergency plans when governments refused to provide them. Why not prepare plans for Massachusetts, or update those previously drafted, in case the petition to shrink the zone were denied? If Yankee were to submit plans for Massachusetts and thereby double the legal avenues toward operation, the company should act quickly: "the faster we get that one on the table, . . . the better off we all are." William Derrickson, the Yankee executive in charge of licensing, asked if the preparation of Massachusetts plans would speed the process. "I think clearly," Stello replied.

Stello's advice reflected the preliminary assessment of Brookhaven Laboratory that Seabrook Station was something less than the engineering marvel its owners claimed. Federal research was incomplete, but "based on the work we have done so far," an NRC staff member reported

to NHY management, "We . . . don't have the confidence your sub-
mittal says you have" in the plant's ability to survive various failures. In
addition to specific misgivings about Yankee's risk assessment study, the
NRC clearly preferred, as a matter of policy, to change the rule rather
than exempt a single licensee from its requirements. From the outset,
members of the regulatory staff had said it would be "difficult" to sustain
claims that Seabrook was uniquely safe; if the licensing board accepted
that proposition, one staff member warned, there would be implications
for several other pending cases, and licensing might become unmanage-
ably chaotic. Yankee's determination to press its petition in the face of
discouraging signals from the agency was a measure of the company's
plight. But the effort was in vain. In April, Judge Hoyt found no basis to
proceed with Yankee's elaborate and expensive petition; in May, Derrickson pronounced it "dead."[54]

While management, politicians, and the upper echelons of the NRC
sought short cuts and revised rules, the routine licensing process, step by
faltering legal step, lurched on. After hearings in the fall of 1986, during
which Portsmouth audiences had been particularly unruly and the
atmosphere unusually noxious, one of the two Atomic Safety and
Licensing Boards dealing with Seabrook Station had authorized operation at "zero power." Two weeks later, after a predictable blizzard of legal
motions had been cleared, nuclear fuel was lowered into the reactor and
testing began.[55]

Intervenors had maintained that introducing radiation, at whatever
concentration, into a plant that might never operate made no economic
sense and might preclude conversion to some other fuel. The argument
rested not only on Massachusetts' refusal to file an emergency plan but
also on alleged inadequacies in New Hampshire's plans, which had yet
to be adjudicated. Judge Helen Hoyt curtailed schedules, issued rulings,
and gave every outward sign that she expected to deal with New Hampshire's plans at a pace so quick even the NRC staff thought it unreasonable. No one—least of all intervenors who had tangled with Judge Hoyt
several times before—doubted her determination, but she underestimated the legal ingenuity of counsel in the case and overestimated her
own authority and efficiency.

She could not even control the calendar. In March 1987, intervenors
asked the appeal board to set aside a schedule Hoyt had ordered. Her
timetable was so constricted as to deprive intervenors of due process,
Robert Backus protested, and was established because of Judge Hoyt's
perception that she needed a secure federal facility to prevent distur-

bances like those that had made headlines the previous fall. The notion was incorrect, Backus said, and irrelevant as well, because due-process rights could not depend on access to a particular building. Further, the disorder Hoyt feared stemmed from the public's view that "this agency is not a neutral and disinterested tribunal but an advocate for the applicants," whose hearings too often seemed "a fraud and a sham." Finally, Backus noted, Judge Hoyt failed to meet her own deadlines and had not issued full opinions on matters that were supposed to be litigated in a few weeks.[56]

The appeal board seemed sympathetic. Judge Hoyt, of course, did not appear before it, but Alan Rosenthal asked Thomas Dignan and the NRC staff "to tell us just what is the deadly rush." "We are going to want to know," Rosenthal predicted, "just what possible justification there was for this compressed schedule," given the inevitable delay before the "plant could conceivably reach a full power license." In "the real world," he continued in a colloquy with Dignan, "the chances of your getting a full power license in the relatively near future are extraordinarily dim." The case had lingered for years, but now, "all of a sudden," Hoyt had promulgated a schedule that seemed "extraordinarily," perhaps "unreasonably tight."

> And I say to myself why . . . are these people put to the task of jumping through this kind of hoop, particularly when . . . the Licensing Board does not meet its own schedule?

Delay, after all, was not the intervenors' fault, Rosenthal said. Other parties—in particular, New Hampshire and Hoyt's own board—had adapted schedules to their own convenience and held up proceedings for years; now they appeared to hurry their opponents unfairly.

Both Rosenthal and Gary Edles took up the question of fairness with Gregory Berry of the NRC staff. Was it fair, they asked, to give intervenors no opportunity to review FEMA's findings and perhaps secure testimony to rebut them, steps that Hoyt's schedule precluded? Berry thought so; intervenors could develop their case through cross-examination. But, the judges persisted, they cannot file rebuttal testimony, can they? Berry deferred to Dignan, who said they could secure rebuttal testimony in advance, a lame answer that would have required a guess about what was to be refuted. Rosenthal expected the NRC staff, which presumably had "no axe to grind," to be "particularly sensitive to the issue of fairness." Berry said the staff in fact was so. Edles had heard enough: "I don't see it quite frankly."

Rosenthal shifted from justice to law. As he read it, submission of rebuttal testimony was a legal right as well as a matter of fairness. Berry cited a case he thought contradicted the judge's interpretation, and Rosenthal brushed the citation aside as not in any way applicable. Could Berry explain to the appeal board, or justify, Hoyt's ruling that rebuttal testimony from intervenor witnesses was not necessary? "Mr. Chairman," Berry said weakly, "I didn't write that footnote."[57]

The order came down a week later, and it was an unusually stiff rebuke. Hoyt's schedule had not provided intervenors "a fair opportunity to prepare for trial," and there was no justification for "such severe curtailment of . . . procedural rights." The failure to litigate New Hampshire's emergency plans in 1983 "or, for that matter, in 1984, 1985, or 1986" was not the result of "foot-dragging on the part of intervenors." The Hoyt board had "manifested an arbitrary unwillingness to make any adjustments . . . even when its own failure to meet the established deadlines" compressed the schedule. And the unwillingness to grant the opportunity to seek and submit rebuttal testimony "patently and seriously intrudes upon the intervenors' hearing rights." Three days later, Hoyt moved the hearings from early summer to mid-September 1987. Just before they were to open, pleading ill health, she resigned from the Seabrook licensing board.[58]

6 | Conclusion

"A $5 billion mess."

By turns irritated and reflective, Robert Harrison reviewed more than thirty years with the Public Service Company of New Hampshire (PSNH) in September 1988. He had resigned as president, a press release said, because of two coronary bypass operations, a history of cardiac disorders, and the advice of his physician to cease his professional activities. The board had elected John Duffett, the company's executive vice-president and chief operating officer, to replace Harrison and named William Scharffenberger, a West Virginia executive, to the vacant post of chairman. The new chairman's qualifications seemed peculiarly relevant; he had had "extensive experience in corporate reorganizations," the utility said, and was the president of another corporation that had been operating since 1985 under Chapter 11 of the bankruptcy code. That provision allows an insolvent debtor, supervised by a bankruptcy court, to do business as usual while negotiating with creditors for a permanent solution to financial distress. For years, Robert Harrison's ability to borrow money had enabled PSNH to avoid bankruptcy, but the Wall Street well had run dry in mid-1987; at the end of January 1988, PSNH became the first public utility in more than fifty years to declare bankruptcy. His varied and ingenious financial schemes had sufficed only temporarily to meet the unending costs of Seabrook Station.

Would he do anything differently, a reporter asked. Maybe, Harrison joked, he should have "gone back to Oklahoma after I got out of school." What about his subsequent decisions about the nuclear plant? In general, Harrison thought the company's efforts had been prudent and defensible, however inept they might seem in retrospect. Management had erred in its forecast of oil prices, interest rates, and construction costs and schedules, he conceded, but those estimates had rested on the best information available at the time, which had always shown completion of Seabrook Station to be the least expensive option open to the company. Were it not for accidents at Three Mile Island and Chernobyl, events obviously beyond PSNH's control, "Seabrook would be operating," Harrison asserted; before Chernobyl, Seabrook's officials had been "very close" to an agreement with Massachusetts that would have removed the emergency planning barriers to a license.

Probably the company had paid too little attention to "what's on the minds of people," Harrison remarked in an oblique concession that PSNH had misjudged the hostility Seabrook Station had provoked. Demonstrations in the 1970s had "politicized" the issue and "raise[d] the stakes," Harrison knew, as well as "our blood pressure or whatever you want to call it." Had those demonstrations hardened management's resolve to complete the project? "I don't know," Harrison replied, but he did know that there would have been considerable "political fallout if we had announced a decision to cancel." Still, "knowing what I know now," should the company have undertaken Seabrook? "Of course not! . . . how could you ask? We're bankrupt."[1]

It had been coming for some time. Reorganization and a large infusion of borrowed money had provided a temporary reprieve in 1984. The company's auditors noted in 1985 that the balance sheet carried the second reactor as an asset, which in fact it was not since expenditures for unproductive facilities could not be recovered. Further, PSNH's earnings for 1985 consisted entirely of what accountants called "allowance for funds used during construction," a noncash bookkeeping entry that paid no bills. Thus, Harrison had written in his annual report to stockholders, who had received no dividends since 1984, the company would have to borrow more than $850 million between 1986 and 1990 to pay the continuing costs of Seabrook and to service the company's debt. In 1986 New Hampshire's Public Utilities Commission (PUC) had explicitly decided that the bankruptcy of PSNH, a distinct possibility, served no one's interest. But, Harrison warned, without more borrowed money, the company would run out of cash after June 1987.[2]

That was more accurate than others of Harrison's forecasts; PSNH missed its first interest payment in September 1987. Early in the year, PSNH had asked the Public Utilities Commission for permission to sell $240 million in long-term debt, of which a portion was earmarked for the continuing expense of Seabrook Station. Pending before the commission at the time was the first of a series of rate increases that PSNH had proposed to ease the eventual shock of adding Seabrook to the rate base. This program—dubbed Pathway 2000 by the company's publicists—projected annual increases that led to doubled rates by 1992, a result that would have contributed mightily to the company's financial recovery. But the PUC denied all but 5 percent of the requested 14 percent increase and, for good measure, entertained the arguments of Seabrook opponents that the company's financial scheme sought concealed charges for construction work in progress (CWIP), which the legislature had prohibited. PSNH withdrew the request for long-term financing and secured instead permission to borrow $150 million in short-term notes to cover expenses unrelated to Seabrook.

But Merrill Lynch, the Wall Street firm that had guided PSNH through the 1984 reorganization and marketed the company's securities thereafter, confessed that a new $150 million borrowing could not be sold on PSNH's terms—4 percent above the prime interest rate. Eventually, Drexel Burnham Lambert placed $100 million at ruinous rates: 9 percent above prime and an additional 0.25 percent for every month the notes remained unpaid. In July, Harrison formally warned the Securities and Exchange Commission that recent events made bankruptcy seem increasingly likely. The Federal Emergency Management Agency (FEMA) had decided that New Hampshire's emergency plans were incomplete; the Nuclear Regulatory Commission (NRC) had delayed a low-power authorization for Seabrook Station because the utility's emergency plans for Massachusetts were an inadequate substitute for plans the governor refused to submit. Without operating funds, which the PUC had denied; without the prospect of a prompt license for Seabrook and thus a return on an investment that constituted nearly 70 percent of the company's assets; without the hope of borrowing money, even at usurious rates, "the company's management and financial advisers have concluded," the solemn announcement continued, "that . . . financings in the amounts projected to meet the company's cash needs during the next several years" will not be available.[3]

New Hampshire Governor John Sununu seemed unruffled. The company and its advisers, he said, were at work on a "financial restructuring package" that would avoid a formal bankruptcy filing. But the "restructuring package," which cost PSNH $4 million in fees, never had a chance of success. Announced in September, the scheme invited bondholders to lengthen the maturity of their loans, to receive common stock and warrants instead of interest payments, and thereby to reduce the company's immediate need for cash. The plan, PSNH's financial vice-president said, "gradually gives the company to the debt holders." But one of the largest of them retorted immediately that he did not want it; the company's plan, said Martin Whitman, who controlled enough bonds to defeat it and who wanted cash instead of the financial equivalent of promises, was "dead in the water." And indeed it was.[4]

Meanwhile, the company maintained that restructuring was only one element in a comprehensive attempt to regain financial stability. In addition, PSNH would reduce expenses, making major cuts in maintenance, payroll, and promotion, though the "aggressive advertising campaign" to persuade New Hampshire of the virtues of Seabrook Station would be continued. The company would not promise to serve new customers, since new services might require new investment. And finally, PSNH returned to the PUC with two requests: an immediate rate increase of 15 percent; and a formal inquiry about the constitutionality of the law prohibiting inclusion of CWIP in the company's rate base.[5]

Rearrangement of corporate debt was not a topic that engaged the public as did rising rates, reliable service, and economic growth for which electricity at a reasonable cost was crucial. Max Hugel, owner of New Hampshire's major racetrack and a Republican well connected in the party's conservative wing, had heard enough. He set up a public address system outside PSNH's corporate headquarters and held a press conference. The Seabrook project, Hugel said, was a "monument to stupidity, arrogance, and mismanagement, . . . a $5 billion mess." When pressed, Hugel said he did not know whether the plant was safe or what ultimately ought to be done with it. But his statement implied reservations about safety as well as outrage at the prospect of skyrocketing rates:

> People . . . should not have to live in fear of higher and higher rates which ultimately could leave them jobless and out in the cold. They should not be left . . . in an environment that is unsafe to live in and which is dangerous to their children's health.

The utility's public relations staff correctly pointed out that Hugel's expertise on these topics had not been established. But other prominent Republicans too began to disassociate themselves from PSNH and what Hugel called "this albatross" at Seabrook. Two leaders of the Republican majority in the state Senate called for Harrison's resignation, and even Governor Sununu, while reaffirming his support for nuclear power, said he was no "ardent supporter of Public Service of New Hampshire." Criticism from Massachusetts Democrats was to be expected—the press secretary of Governor Michael Dukakis disparaged the plant as a "colossal financial blunder" and the "bailout plan" as a "white elephant"—but softening support among New Hampshire Republicans was not a good omen.[6]

Paul McEachern, the Democrats' candidate against Sununu in 1986 and the attorney for Hampton in the licensing proceedings, turned the headlines into a petition to the Atomic Safety and Licensing Board (ASLB). McEachern knew that the Nuclear Regulatory Commission had eliminated a requirement that utilities owning and operating nuclear power plants be "financially qualified." But the rule reserved authority to investigate "special circumstances," and surely, McEachern wrote, PSNH's circumstances were now special. A low-power authorization and other licensing questions ought to be delayed, he held, pending determination that a corporation in desperate financial circumstances would protect the public. The ASLB routinely denied his petition because McEachern had failed to demonstrate that PSNH would not eventually recover its costs for Seabrook and thereby be financially qualified when operation began. Other intervenors joined Hampton in a prompt appeal, which was still pending in January 1988, when PSNH formally declared bankruptcy.

Although the licensing board declined to investigate PSNH's financial condition, the NRC did ask Robert Harrison several questions. He reiterated PSNH's determination "to successfully complete the licensing process." Even if bankruptcy were to occur, Harrison said, revenue from continuing electrical sales would cover PSNH's share of Seabrook's costs; indeed, he noted, bankruptcy would ease the problem by temporarily postponing large interest payments. Harrison did not say that the company expected to omit interest payments due in October, taking a calculated risk that creditors would not push the company into insolvency. Rather, he expressed confidence that his restructuring scheme—the one Hugel had criticized and bondholders killed—would enable PSNH to return to financial health.[7]

Meanwhile, the company's directors, as New Hampshire papers noted pointedly, took care of themselves. The board paid itself $150,000 in fees and awarded "golden parachutes" to managers, promising three years' compensation even if they were fired or the company went bankrupt or was taken over. "The peasants in the company," the *Hampton Union* observed, received no handsome guarantee, a situation like that of "the peasants out in the countryside trying to pay their electric bills."

As the company slides toward insolvency, we find the top people . . .
thumbing their noses at the public. . . . The company may be going bankrupt, the ratepayers may be asking where their electricity is coming from and
what it is going to cost, but the top folks don't have to worry about those
matters. No, they're too busy getting theirs before it's too late. . . . No doubt
this is all legal. It just isn't right, that's all.[8]

Given a pecuniary vote of confidence, management tried to keep the
corporation afloat. Before the end of 1987, however, bondholders indicated impatience, and on January 27, 1988, the New Hampshire Supreme Court pushed PSNH toward the shelter of the bankruptcy court.
The state's prohibition of rates based on construction work in progress,
ruled New Hampshire's highest tribunal, was constitutional; the emergency rate increase that was crucial to PSNH's scheme was, therefore,
dead. Investment, even in regulated utilities, the justices held, carried
risk, and investors, "not ratepayers," must bear the consequences of that
risk; the "bailout" the company sought was denied. An earlier effort—
the fourth—to secure repeal of the anti-CWIP law had failed in the
legislature, and an appeal to the Supreme Court of the United States,
which was ultimately unsuccessful, would require months.

PSNH had only hours. Two days after the decision of the New
Hampshire court, the company filed for protection under Chapter 11 of
the bankruptcy code. The event had been so long anticipated that it
was almost anticlimactic. Corporate employees, Robert Harrison said,
would not be paid for a couple of weeks, but otherwise it would be
business as usual. Investment bankers, asked to comment, thought the
decision inevitable and probably beneficial because the company had
secured "some breathing room," as a specialist in the securities of bankrupt companies remarked. Governor Sununu's sour statement blamed
intervenors and indecisive management for the debacle.[9]

"I don't think this will be a factor in the plant's licensing," remarked
John Eichorn, chairman of Eastern Utilities Associates and New Hampshire Yankee, the consortium that owned Seabrook Station. But intervenors, already appealing a ruling that barred the reopened inquiry into
PSNH's financial qualification, fired a new salvo of legal motions to the
NRC. Bankruptcy, *"in and of itself,"* was enough, Robert Backus argued, to require a full-scale investigation of PSNH's financial status.
The NRC's rule assumed that a state's rate-setting procedure assured the
stability of a regulated utility. Surely, intervenors held, the uncertainty
attending a bankruptcy filing undermined that postulate. To clinch the

case, Backus quoted PSNH's financial vice-president, who admitted that "nothing . . . in bankruptcy" was "a certainty."

The NRC's legal staff had anticipated Backus's point. In November, when bankruptcy was still only a probability, Gregory Berry told the appeal board that a formal plea was not a "special circumstance" that would require reexamination of PSNH's financial qualification. Eventual operation, Berry contended, would enable the corporation to recover its investment and restore financial respectability. Intervenors challenged the objectivity of a regulatory agency that assumed the utility would prevail in still unfinished adjudicatory proceedings while opponents were forbidden to assume the converse. But intervenors had long since lost any illusion about the NRC's objectivity, and the commission had already ruled in the Shoreham case that doubts about eventual operation would not block a low-power license, which was technically the disputed topic at Seabrook.[10]

The appeal board, aware of PSNH's peril, invited the views of participants in the case. To the argument that PSNH's partners would undertake a larger share of Seabrook's continuing cost, Attorney General James Shannon of Massachusetts pointed out that those companies were not immune to the financial virus that had disabled PSNH. Several of the consortium's smaller partners had declined to pay their share of Seabrook's costs for some months. United Illuminating, the second-largest participant, told Shannon that it could invest no more in Seabrook without the approval of Connecticut regulators, who had already concluded that about a quarter of the funds expended there were imprudently invested. A subsidiary of Eastern Utilities, formed to purchase (at a substantial discount) the shares of partners that had decided earlier to abandon the project, had run out of cash; without new financing, the company would soon default on its own interest payments. The Massachusetts Municipal Wholesale Electric Company, Seabrook's fourth-largest owner and the representative of several municipal utilities with small shares of Seabrook Station, was being urged to divest by managers of member companies, who in turn were under political pressure from their constituents. None of the owners of Seabrook Station, Shannon held, was financially robust, and the debilitating cause was not difficult to diagnose.

The appeal board conceded that the "unprecedented" bankruptcy filing gave intervenors' arguments "a certain visceral attraction," which the judges ultimately resisted. That panel agreed with Gregory Berry that intervenors had not met the legal test for reopening questions about financial qualification. But the board did not close every door: It asked the NRC to decide whether Attorney General Shannon had made a case that the joint owners of Seabrook Station, given their collective financial distress, might be unable safely to conduct low-power testing. To that issue, the commission itself added a query about funds to pay for safely decommissioning the plant. [11]

The fact that the NRC kept the issue alive may have worried Seabrook's owners, because they asked Judge James Yacos, who was presiding in the bankruptcy case, to permit New Hampshire Yankee to become independent of PSNH. What, Judge Yacos asked, "is so important about doing that now?" PSNH's attorneys thought that dividing the two entities might speed the licensing process; probably association with a bankrupt lead partner was no advantage. Intervenors pointed out that a separate New Hampshire Yankee would diminish the judge's authority over Seabrook-related expenditures. He would not, for instance, be able to bar PSNH from undertaking the costs of low-power testing. Yacos decided to keep his options open and denied the request. [12]

For its part, in December 1988, the NRC concluded that low-power testing did not present a safety hazard. The opinion was carefully hedged; the commission explicitly did not hold that the owners of Seabrook Station were financially qualified; rather, "on the facts of the case," the commission ruled that "lack of financial qualifications does not pose a significant safety problem." The commission did not accept the owners' estimate of the potential cost of decommissioning and instead calculated itself that $72.1 million would do. If that sum were guaranteed by financially secure members of the ownership group, or by insurance companies, the financial barriers to low-power testing disappeared. [13] The owners and their insurers quickly gathered the requisite guarantee.

A euphoric vice-president of New Hampshire Yankee predicted that low-power testing would begin within a month; he ought to have known better. Emergency planning and other unfinished regulatory business languished in the NRC's judicial hierarchy, and any of a half-dozen potential appeals might produce further delay. Incoming New Hampshire governor Judd Gregg had said that regulatory questions ought to be answered before low-power testing began. And Lando Zech, chairman of the NRC, had twice remarked that financial stringency caused him more unease than it seemed to stir among Seabrook's owners. He would review the matter again, Zech said, when they asked to operate the reactor commercially.

Robert Backus picked up that cue. Though the NRC had held that low-power testing posed no hazard, Backus admitted, full-power operation would undeniably create significant risk. Seabrook's financially strapped owners might conceivably subordinate safety to profit, and thus, Backus argued in yet another brief to the appeal board, the NRC ought to face and resolve the issue. He anticipated the rejoinder that commercial operation would allow inclusion of Seabrook's costs in PSNH's rate base, thereby boosting the company's revenues. But, Backus pointed out, several reorganization proposals explicitly removed Seabrook Station from PSNH's assets, in which case whoever wound up owning the facility might not be a regulated utility, would not necessarily have undergone the NRC's regulatory scrutiny, and might not be financially qualified to operate a nuclear generating station. In any case, Backus held, licensees were supposed to meet regulatory requirements before operation, not afterward, and the NRC ought not to presume in advance that a commercial license would be a financial panacea.[14]

Although the NRC had concluded, at least temporarily, that a company's financial resources did not affect public safety, those resources were clearly of first concern to the bankruptcy court. During 1988, while the NRC disposed of—intervenors said "evaded"—the issue, Judge Yacos kept PSNH, its creditors, and public officials talking about an end to the fourth-largest bankruptcy proceeding on record. He knew the case was extraordinarily complex. There were fourteen outstanding equity issues, for instance, and twenty-four different categories of bonds and debentures. PSNH was defending legal actions brought by partners alleging mismanagement in Seabrook's construction, and by stockholders and creditors. In addition, several regulatory agencies had an interest in the case: the Securities and Exchange Commission, which had jurisdiction over stocks and bonds; the Federal Energy Regulatory Commission (FERC), which dealt with wholesale sales of electricity; the New Hampshire Public Utilities Commission, which would resist any attempt to preempt its control of retail rates; similar commissions in other New England states, where PSNH's partners did business; and the NRC. "In a real sense," Judge Yacos observed, "this case is unprecedented."

It was also expensive. Judge Yacos's conservative estimate was "in the vicinity of $350,000 per day" to pay lawyers, investment bankers, and other expenses that did not include interest lenders could not collect. (The court's New England frugality was manifest when claims by PSNH's attorneys for first-class coast-to-coast air fare were denied.) Although Judge Yacos knew that the presidential election and other events over which he had no control might affect the litigation, he wanted all those expensive experts to keep unremittingly to their work. " 'Waiting around to see what happens at Seabrook,' " he wrote, "is not a sufficient explanation for non-results by professionals paid at the compensation levels authorized by this court's order. . . . If the response is to be 'we can't get there from here,' we probably can find professionals at much lower rates who can tell us that."

But "waiting around" was exactly what PSNH had in mind. "Neither the company nor the professionals" who were collecting large fees had "an incentive to settle," remarked Martin Whitman, a bondholder who did have a multimillion-dollar incentive to settle. Professionals, Whitman charged, had "a cow that can be milked," and the company hoped "to hose the ratepayers." The situation might have been more elegantly characterized, but in fact the company was in no rush to discard the flexibility and protection whose high cost Judge Yacos deplored. He had the authority, PSNH's attorneys asserted, to cancel contracts the company had made to purchase electricity at high prices from small producers, and other agreements undertaking to sell it at low rates to other utilities. Bankruptcy might also divert potentially expensive claims by Seabrook partners that construction had been ineptly and wastefully managed. Whatever the monetary and public relations costs, in other words, bankruptcy seemed to make possible exemption from burdensome regulations, past managerial mistakes, and the oversight of unsympathetic agencies.[15]

In particular, according to PSNH's attorneys, Judge Yacos could let their client out of the box created by the anti-CWIP law and the New Hampshire Public Utilities Commission. In March 1989, PSNH asked Yacos to approve a complex reorganization that established a generating company (Genco), a distribution company (Disco), and a service company (Servco), all of which would be legally independent but which would be owned by PSNH and linked by contracts. Genco would provide wholesale electricity to Disco, which would sell power to retail customers. The Federal Energy Regulatory Commission regulated wholesale rates and, as an increasingly informed population in New Hampshire realized, permitted CWIP to be included in the rate base. ("FERCed again" was a wry New Hampshire reaction to this development.) Although the Public Utilities Commission would continue to have jurisdiction over Disco's charges to customers, those rates would reflect Genco's high costs, including the cost of Seabrook Station. The scheme would probably result in higher rates than PSNH had proposed before entering bankruptcy—rates that the state agency had turned down.

Chapter 11 provided not only shelter and time but also leverage. Bankruptcy proceedings, after all, are supposed to maximize payments to creditors, a role that was in conflict with the economic interest of customers who lived and voted in New Hampshire. PSNH used the possibility that the state might lose regulatory control of rates to pressure political leaders to make a deal. The company said it preferred agreement with the state to more drastic reorganization. But not on any terms; PSNH needed "appropriate and necessary rate relief," John Duffett said, which had not been forthcoming. The state's rejoinder, leaked to the public, was that the company enjoyed a monopoly that the state could revoke or perhaps take over. Members of the state's executive council talked about a state power authority that might operate generating facilities and sell electricity to customers—"a hostile takeover by the people of New Hampshire," quipped a Republican aide planning the statehouse transition from Sununu to Judd Gregg.

Sununu himself, about to leave to become chief of President George Bush's White House staff, thought the prospect of federal regulation "so bad" that even a state takeover, however intrinsically distasteful, was preferable. Republican leaders of both legislative branches shared Sununu's outlook: "I've fought for Seabrook for seven years because I felt it would be beneficial for the ratepayers," said Senate president William Bartlett.

> I didn't do it so PSNH investors would reap big profits. . . . Those investors knew the risk they were taking and now after most of them have got their money back they are trying to avoid risk and maximize their investment.[16]

Behind the game of chicken being played in the press, officials of the state and the utility tried to compromise. PSNH asked for rate increases totaling between 30 and 40 percent; the state countered with 4 percent annually for five years. No, John Duffett replied; "PSNH cannot allow its investors' contributions to this state's economy to be impaired by an apparent political unwillingness to . . . repay what PSNH, with the state's approval at every step, had to borrow to provide electricity." The company wanted the state to allow costs of future construction to be "passed on to ratepayers before the projects operate," to reduce rates "small power-producers can charge" PSNH for electricity it was required to purchase, and to enable the utility to develop Seabrook's abandoned second reactor "as a non-nuclear power plant."

Larry Smuckler, the assistant attorney general who was conducting negotiations for the state, vainly searched Duffett's letter for a concession; perhaps Duffett's tone was conciliatory, Smuckler said, but the substance constituted a declaration of war. The market simply would not bear the monopolistic rates PSNH proposed, Smuckler asserted, and Chapter 11 "did not repeal the laws of economics." As if to emphasize the state's position, the PUC opened an inquiry into what appeared to be windfall profits that PSNH derived from protection from creditors and surging power sales. The chairman of the PUC did not promise ratepayers a refund, but he did hold out the possibility that his colleagues might order reduced rates.[17]

An injunction from Judge Yacos put a quick end to that prospect. He also warned that his tolerance of delay was approaching a limit, and he urged the utility to conduct negotiations swiftly with the state and other parties. Otherwise, he said, the court would welcome proposals from creditors and other corporations interested in a merger, reorganization, or purchase of portions of the company. Two utilities had made public bids for PSNH, and several others were reportedly at work on competitive offers. Both Northeast Utilities and New England Electric System, which had merger propositions on the table, promised to protect ratepayers from the fearsome increases that seemed in prospect if Seabrook began operation. Northeast Utilities proposed to absorb the financially profitable part of the business and spin Seabrook Station off in a new, independent corporation owned by the holders of PSNH's unsecured debt, an action that might present the NRC with a dilemma, since the new entity would have had no regulatory scrutiny. New England Electric, already a participant in Seabrook and distributing electricity to customers in New Hampshire through other subsidiaries, advanced a less drastic division of PSNH's assets.[18] For creditors and investment bankers, the offers may have differed by about $500 million, but for consumers there appeared little reason for preference.

Although a bill enabling the state to take over New Hampshire's largest utility bogged down in the legislature, Governor Judd Gregg retained considerable influence through legal maneuvers in the bankruptcy court. Gregg asked Yacos to block low-power testing, which would effectively interrupt the NRC's procedure for granting an operating license. His action, the governor said, was an effort to spare customers perhaps $25 million in potential cleanup costs if the reactor was irradiated. That sum, of course, seemed inconsequential when compared to what PSNH's customers would pay if Seabrook Station became operational. The company's lawyers promptly offered a legal promise not to seek reimbursement for those costs, and Gregg dropped his objection. Judge Yacos then informed the NRC that he no longer opposed low-power testing, which was promptly authorized. A disgusted Robert Backus charged that the governor had given a substantial tactical advantage away in return for "an after dinner mint."[19]

About the time PSNH skipped its first interest payments, the Atomic Safety and Licensing Board reopened its investigation of emergency planning. The timing was coincidental. Judge Helen Hoyt's last-minute indisposition led to a brief delay while the board was reconstituted under the leadership of Ivan Smith, who needed time to become conversant with tens of thousands of pages of record. More relaxed and less provocative than Judge Hoyt, Smith convened the hearings early in October at the statehouse in Concord. They began with tight security, street theater, and disruption, and Helen Hoyt would not have enjoyed them at all. Suspicious, inept guards helped raise the level of irritation by fumbling examinations of television equipment and the briefcases of lawyers. Spectators unfurled a banner inscribed "People, Not Profit," and occasionally contradicted witnesses by holding up signs bearing the single word "NO." Some wore gags to protest the board's ruling that members of the public would not be permitted to speak; the board had several volumes of testimony given on other occasions, Judge Ivan Smith ruled, and that evidence of the public's attitude would suffice.

In fact, Judge Smith and his colleagues did not listen to members of the audience that Monday afternoon, but other people did. Smith left the room during several unruly recesses that took the form of an unofficial New England town meeting. Congressman Bob Smith scurried between NRC functionaries and attorneys for the intervenors, who in turn talked with Roy Morrison, Diane Dunfey, and others among the vocal group. Eventually, a compromise permitted "local officials" to make brief, "limited appearance" statements to the board. The definition of "local official" proved elastic enough to accommodate not only Rennie Cushing and Beverly Hollingworth, who were members of the state legislature, but also Guy Chichester and Stephen Comley, who had no claim to official status, both of whom received prolonged ovations from the audience.[20]

The subject of those speeches and the hearings they interrupted was the thirty-one-volume set entitled New Hampshire Radiological Emergency Response Plan (NHRERP), which the state said would protect the public in the event of a nuclear accident at Seabrook Station. Years in gestation, sometimes without consultation either with local officials who would have to implement the plans or with the citizenry they were supposed to protect, the volumes created a buzzing interest when they were deposited in local libraries along the New Hampshire seacoast. Antinuclear volunteers, using knowledge acquired in a lifetime of residence, pored through the volumes noting discrepancies, omissions, lack of familiarity with traffic patterns, and mentally measuring local officials against the tasks they were assigned. The result of this critical research piled up in the offices of Paul McEachern and Matthew Brock, who represented Hampton in the NRC's proceedings, or Robert Backus, attorney for other seacoast towns and for the Seacoast Anti-Pollution League (SAPL). The critique was detailed and, from the point of view of those who compiled it, devastating, because it revealed the unfamiliarity of planners with local conditions and the inability of the specific plan— many believed of any plan—adequately to protect people in the region.

The growing corps of attorneys in the case eventually agreed to litigate objections in eight segments: letters of agreement, which the state was supposed to secure from those who were to supply services such as ambulances and tow trucks; the availability of people to carry out traffic control, drive buses, decontaminate vehicles, and the like; the adequacy of transportation arrangements, especially for hospital patients, schoolchildren, and others unable to drive themselves; preparation and staffing at reception centers, where evacuees would be decontaminated and, if necessary, provided temporary residence; the system of sirens and radio broadcasts to alert people to an emergency, and the ability of the telephone network to meet the demands an emergency would impose; conflicting views about the behavior of people in a radiological crisis, which varied from a predicted "therapeutic community," in which people would behave altruistically, to vows to take care of self and family before tending to public duty; the existence of useful shelter, particularly for the transient population at the region's beaches, an inquiry that led to an inconclusive examination of political influence on the licensing process; and the disparate studies of the time required to evacuate when automobiles might skid or overheat or collide, or when impatient drivers simply decided to walk away from a nuclear disaster.[21]

During the first afternoon of the hearings, in a statement rowdy spectators would not let him complete, Thomas Dignan, who led a covey of lawyers for Seabrook's owners, outlined his clients' legal position. The "reasonable assurance" that Judge Smith and his colleagues had to find before the plant could operate, Dignan said, was not the same as "absolute assurance of 'zero risk.'" The NRC's rules did not require the company or the state to provide new facilities to guard against every imaginable accident, nor was there any defined level of radiation exposure from which the public must be shielded. He did not, for this hostile audience, develop the logical and legal consequences of this reading of the applicable law, though he would, of course, do so as the case progressed.[22]

Fundamentally, Dignan's case boiled down to what Massachusetts Attorney General James Shannon called a "best effort" interpretation of the regulations, while Shannon himself held that a "best effort" might still fall short of the required "reasonable assurance that adequate protective measures can and will be taken" to assure public safety. Shannon maintained, in other words, that plans could fail to meet an absolute standard that reasonable people would agree was appropriate to protect those whom the plant put at risk. His test involved judgment—What constituted enough protection, and what level of assurance was reasonable?—and implied a conclusion based upon investigation of local circumstance. Dignan's test, on the other hand, assumed that no emergency planning shortcoming could block a license. Planners must try to make local peculiarities fit federal regulations, and that effort, by definition, sufficed. If, in fact, people on the beach were not and could not be protected, that was, in Dignan's insouciant phrase to the Seabrook appeal board, "show biz."

The phrase implied that, whatever the problems of an unprotected public, they were not his, which was not quite accurate. Unless the ASLB could conclude that indeed there was "reasonable assurance," Dignan's clients would get no license. But he argued that *how* safety was secured was not the responsibility of Seabrook's owners. Federal regulators at the NRC and at FEMA and civil defense officials of the state of New Hampshire would make the decisions; PSNH and the other utilities would pay most of the bills, which would ultimately be charged to ratepayers who, in Dignan's view, were unduly anxious about a nuclear emergency anyhow. When FEMA threatened to block a license, Dignan injected himself decisively into the proceedings. Otherwise, he was semidetached, bantering with judges and other counsel, and conveying a serene confidence that New Hampshire's splendid plans would carry the day.

Even the intervenors were doing him a favor. The flaws they had discovered in earlier editions of the NHRERP had led to revisions that improved the final result. Where new investigation turned up additional imperfections, they too would be corrected, because emergency plans never assumed a final form. If more training was necessary, it would be given; if more expense had to be incurred, those who demanded the expenditure would finally pay for it. If Dignan's reading of the law was correct, he needed only to be patient; he could not lose.

Intervenors saw both the law and the outcome differently. The Commonwealth of Massachusetts gathered expert testimony to contradict the utilities' evidence about radiation exposure and evacuation time and the availability of personnel to manage the departure of vehicles and people from an area served by inadequate roads. For technical reasons, Smith did not admit all the material intervenors collected. But if their case did not invariably follow their own script, they did nevertheless present it without the crackling hostility that had characterized their efforts when Judge Hoyt presided.

The central thesis of their case was the familiar assertion that for any of dozens of reasons the plans were inoperable. If any single flaw seemed trivial, in the aggregate they demonstrated conclusively that Governor Michael Dukakis of Massachusetts had the right idea: A nuclear accident in that location endangered too many lives. If everything went as New Hampshire planners hoped, on a sunny summer weekend eight hours might be required before the last cars could drive away from Hampton Beach. But, intervenors argued, the plans were necessarily so complex that Murphy's law—if something can go wrong, it will— would obtain. State police would have to come from hours away to direct traffic. Members of the National Guard would not be available for the various tasks for which they were, in any case, untrained. Buses with plywood "conversion kits" to fashion into beds were inadequate substitutes for ambulances that would have to converge on Seabrook from communities up to 100 miles distant. Prescripted messages for emergency broadcasts were vague about bus routes, public shelters, and other details, and their bland reassurance would not reduce panic. ("Any release of radiation will not exceed levels recommended by the Environmental Protection Agency," as if the EPA had established levels that were in some way good for health.)[23] Public employees, including teachers, could not be assumed to be available to fulfill roles planners prescribed for them. The state had underestimated the population of the region, the number of cars, the extent of tourism, and the need for public transportation, and overestimated the capacity of roads, the public's patience and docility, and the ability of fallible planners to control a crisis.

Officials from coastal New Hampshire—teachers, police, civil defense directors, selectmen—testified that they could not perform some of the tasks the state assigned and would not perform others. The state counted on employees from Hampton's Public Works Department for various functions, for example, but half of those enumerated were summer employees. The Police Department, whose officers had unanimously denounced the scheme as "unrealistic, unworkable, and unsupportable," was also overcounted. Police in neighboring towns in both states agreed that long experience with summer crowds and the region's roads made them doubt that the NHRERP could work. Local officials from the seacoast reiterated that the state expected more than could be delivered; repeated promises to supply lacking staff and equipment from state resources were discounted by people for whom Concord's unfulfilled promises were entirely too familiar.[24]

A panel of teachers told the ASLB that, instead of monitoring buses and supervising pupils, they would first provide for their families. The local teachers' union formally notified the superintendent of schools that imposing additional duties during a radiological emergency appeared to alter the terms of employment without collective bargaining and therefore to violate the contract. The NHRERP rested on "dangerously unrealistic" assumptions and contained "obvious inadequacies," and the membership "overwhelmingly rejected" their responsibilities and would refuse to fulfill them. A state court ruled that in fact private citizens, even if they were public employees, could not be required "to carry out the NHRERP in the event of a radiological emergency." Richard Strome, the state official most identified with the plans, backed away from the verb "shall," which recurred in the text; it did not imply a command, he said, but only the state's view of what people ought to do. If Strome's definition obtained, the plans rested on willing service by volunteers, and most of those thirty-one volumes consisted of a fantasy about what might happen if people behaved according to Strome's precepts, with which teachers on the seacoast, among others, explicitly did not concur.[25]

The response to intervenors' criticism was predictable: Plans are not perfect or permanent; defects will be fixed. Creating workable plans, the state noted, was made more difficult by the refusal of Massachusetts and several New Hampshire towns, including Hampton, to cooperate. Echoing Governor Dukakis's assertion, the nonparticipating towns argued that the public's safety could not be assured and that emergency planning in consequence was an empty and ultimately disingenuous gesture. Yet nonparticipation also removed some of the sting from their critique. When town officials refused to provide information about the number of police officers, pupils, or disabled citizens, or declined to furnish data about traffic and parking and shelters, how could those officials reasonably complain when the state made mistakes? Members of several NRC boards had asked that question in the past and expressed bewilderment or irritation at responses that did not explain. Taken on the advice of counsel, the decision not to participate in drafting emergency plans may have blunted the towns' best weapon: their detailed mastery of local circumstance.

Noncooperation was not restricted to planning. Appointed officials, including civil defense directors and police chiefs, like teachers, vowed not to carry out assignments if an accident actually occurred at Seabrook Station. Their refusal was not a capricious dereliction of duty, these witnesses maintained, but stemmed variously from distance, a primary obligation to family, or a principled opposition to nuclear power. The Nuclear Regulatory Commission disparaged such statements as campaign speeches, discounted them, and directed licensing boards to assume public servants would do their civic duty.[26]

Edward Thomas intended to do his. As chief of the Natural and Technological Hazards Division of the Federal Emergency Management Agency in Boston, he was supposed to review New Hampshire's plans, test them in an exercise, and submit an evaluation to Washington. If formally endorsed by FEMA, his assessment would become the basis for the issuance or denial of a license. It was a weighty obligation, shared with representatives of other governmental agencies in a Regional Assistance Committee (RAC) from which Thomas expected to coax unanimity and thus joint responsibility. But FEMA had to make the final judgment, and as FEMA's representative, Edward Thomas was in charge.

He hoped for consensus in part because he knew the political heat that even a scheduling decision could create. Governor John Sununu had tried to accelerate the test of an early version of New Hampshire's plans before FEMA believed them ready. (Sununu said that reports of his intervention were "ridiculous" and "absolutely false," but the president of New Hampshire Yankee conceded that he had asked for the governor's help about the time the schedule was set.) After the exercise, in February 1986, when FEMA identified crippling deficiencies in the plan, Sununu became "wildly upset," according to a congressional staff member, and told people in the FEMA hierarchy he was "on his way to the White House to clean out staff operations." Whether Sununu in fact complained to the White House, and whether or not his complaint made any difference, Thomas certainly knew that FEMA's unfavorable review had caught the governor's disapproving notice. [27]

If Sununu wanted haste in 1986, neither he nor the utility had been in any hurry in 1982 and 1983, when the NRC had asked Thomas for a progress report in order to keep the licensing process moving. He had, at the time, nothing to assess. Later, when Thomas asked questions about buses or drivers or protection of people on the beach, which was the issue that troubled him most persistently, planners brushed his concerns aside "because Seabrook was special." The plant had unusual safety features, Thomas was informed, that had convinced the utility, the state, and the NRC that the usual rules did not apply and that the plant was "entitled to special treatment and . . . consideration."

That was a proposition Thomas was willing to entertain, but he needed a written, technical justification on which to base the exception. In February 1987, Robert Bores, the NRC's representative to Thomas's RAC, advised him that the design of Seabrook's containment would assure even a swollen summer population enough time to evacuate before receiving dangerous levels of radiation. That memorandum was, Thomas recalled, "the solution to the problems that FEMA had been raising concerning the . . . beach." [28]

But the solution vanished. In May, Bores called Thomas to report "horrible things going on" because the "lawyers are involved." Although Bores stood by his memorandum, he warned Thomas that it might be retracted, and the two commiserated about the consequences for Seabrook's license. Whatever Bores's professional judgment about Seabrook's containment, it had been flatly contradicted in April when Helen Hoyt threw out New Hampshire Yankee's petition to shrink the emergency planning zone around Seabrook. In effect, Hoyt had concluded that Seabrook's containment was not unusual enough to justify an exception to regulations, and the plant's champions did not want to give intervenors the opportunity to use her decision to discredit Bores and any assurance of safety that FEMA rested on his professional advice.

Lacking the Bores paper, Thomas was in a quandary. Although the NRC argued that the memorandum was redundant and that the plant met requirements, Thomas continued to believe an accident would pose an unacceptable hazard to unprotected bathers. "Reasonable assurance," he thought, required more than compliance with the letter of regulations, and David McLoughlin, to whom Thomas was responsible, agreed that experience and judgment, as well as a reading of rules, should shape FEMA's eventual finding. Thomas weighed the NRC's argument for a few days and then told the press that the NHRERP did not inspire the requisite "reasonable assurance." The plans still provided inadequate safeguards for summer tourists, Thomas said. His conclusion was not final, and further revision might meet his agency's objections. Perhaps Seabrook Station might be closed in the summer, or shelters could be constructed near the beaches, or he might receive new technical information, Thomas suggested. But when Judge Smith convened the hearings in the fall, Thomas had not changed his mind. [29]

If his testimony stood up, Seabrook Station might never operate, a prospect that ended the effort to achieve consensus in the intergovernmental RAC over which Thomas presided. He knew that the group was deeply, perhaps bitterly, divided over his belief that there must be shelter at the beach. Proponents of the plant did not want to predicate emergency arrangements, and therefore a license to operate, on the availability of protection for transients. Buildings near the beach, which might shield people temporarily, were often flimsy seaside cottages, unwinterized and constructed to take advantage of outside breezes rather than to keep them out. Further, the buildings were privately owned and not automatically available to the general public. A consulting firm had identified structures with sufficient square footage to accommodate the summer population, but planners had not investigated their use. Instead, the state decided to rely on evacuation as the best protective strategy under almost any conceivable circumstance.

New Hampshire's decision did more than make a virtue of perceived necessity. Even the prolonged evacuation of a summer weekend, planners predicted, would expose the public to less radiation than would temporary sheltering and later evacuation when the hazard might well be greater. This response satisfied most federal reviewers, but not Edward Thomas, who presented FEMA's testimony to the Atomic Safety and Licensing Board late in 1987.

Thomas's stance, which appeared to intervenors as the principled defense of his professional assessment, seemed to Seabrook's owners and many of Thomas's governmental colleagues to be a stubborn, almost perverse, refusal to recognize fact and applicable law. Thomas Dignan challenged Thomas's candor, his competence, and his judgment in order to discredit his testimony. More important, Thomas's support in his own agency was eroding. David McLoughlin, a career FEMA functionary who had backed Thomas while acting as his hierarchical superior, was replaced by Grant Peterson, a political appointee, in mid-December 1987. In January 1988, FEMA's counsel in the Smith hearings announced that the agency was reviewing its previous testimony. Meanwhile, Victor Stello, the NRC's blunt executive director of operations, told Peterson that FEMA must not interpret the NRC's regulations, as Edward Thomas appeared to be doing; such interference, Stello warned, would trigger a bureaucratic war. Late in February, members of Thomas's RAC seemed impatient, even irritated, at their chairman's continuing unwillingness to accept New Hampshire's plans. Early in March, after studying New Hampshire's rationale for preferring evacuation over sheltering, a meeting of FEMA officials decided as a matter of agency policy to approve the NHRERP. Thomas said he could support this new position, though he disagreed with it professionally and personally. Peterson decided that it was both "proper" and "caring" to replace Thomas as the FEMA official responsible for the Seabrook proceeding.[30]

But intervenors cried foul. Thomas's resistance to bureaucratic and political pressure seemed to them persuasive evidence of the depth of his conviction, the strength of his character, and the inadequacy of the NHRERP, which he continued to fault. The fact that an officer of New Hampshire Yankee, once employed at the NRC, had warned Thomas months before that he might be relieved of his responsibility for Seabrook seemed to suspicious critics one more clue to the corrupt influence that linked regulators and the nuclear industry. And Peterson, who had removed Thomas and revised FEMA's policy, was a political appointee, perhaps the product of John Sununu's reported vow to clean up FEMA's staff. Peterson's qualifications to assess Seabrook's emergency plans, intervenors noted acidly, included running an appliance store, losing a county election, and managing Ronald Reagan's 1984 campaign in Washington State; that sort of apprenticeship would leave him overmatched in a bureaucratic tussle with Victor Stello.

Edward Thomas was not sure Peterson needed Stello's intervention anyhow. During their first, brief, social encounter, Thomas said, Peterson remarked that flawed emergency plans ought not to idle a $5 billion investment. Later, Peterson told his subordinate pointedly that he grew concerned when "one wagon" was off by itself, "and all the other wagons were over here in a circle."[31] Peterson said the first conversation never occurred and did not relate the metaphorical wagons to Seabrook Station. But Thomas clearly believed he had been given a nudge, and probably an instruction.

Senator Warren Rudman asked Peterson to explain events the press called "FEMA's flip." The agency's decision, Peterson wrote the New Hampshire Republican, had not stemmed from "pressure from the NRC or any other outside group or agency," an assurance that Rudman told reporters he did not believe. Gordon Humphrey, New Hampshire's other Republican senator, who was not ordinarily given to the verbal flourishes of Seabrook's foes, was more expressive: "It smells," said the senator, "like a fish that's been dead for a week."[32]

Senator Humphrey's offended nose, however, was not evidence that might in court seem denial of due process. Even under the relaxed rules of evidence in an NRC hearing, intervenors were not able to prove the political intervention they were convinced had reopened the path to Seabrook's operating license. FEMA's officials portrayed Thomas as a recalcitrant maverick, unable to assimilate evidence that might alter his conclusions, isolated on this issue from the better-informed views of colleagues. To be sure Peterson and others at FEMA talked with Sununu and other political leaders; of course they met with Stello and his associates at the NRC. But such contacts were normal, not corrupting, and conveyed information, not influence. Only occasionally did Peterson or McLoughlin resort to "I do not recall" or some other evasion on the witness stand, and prolonged examination uncovered no "smoking gun," no evidence of improper intervention.

Yet intervention need not be improper to be effective or covert to be heeded. Governor Sununu did not hide his support for Seabrook Station, nor was his prominence in Republican politics unknown. Support for nuclear power in the Reagan administration was no secret. Secretary of Energy John Herrington, for instance, had publicly urged the NRC to permit utilities to devise emergency plans if states refused to do so. In 1988 the president ordered FEMA to circumvent uncooperative state and local governments by preparing federal plans if necessary to move the licensing process forward. When the order was signed, after the 1988 election, Governor Sununu said he had not discussed it for months, but the New England press reported that earlier in the year he had lobbied the White House almost daily. In any case, the order was a clear signal that "the administration supports nuclear power and safe nuclear power plants," as a member of the White House staff put it. That attitude was not likely to change when John Sununu became chief of the presidential staff early in 1989. Nobody had to tell Grant Peterson which way the wind was blowing.[33]

Or his subordinates either. At the end of June 1988, FEMA officials gathered near Seabrook Station to observe a formal exercise of emergency plans covering both New Hampshire and Massachusetts. Although a 1986 test for New Hampshire only had disclosed major shortcomings that escalated Edward Thomas's misgivings and John Sununu's rage, and although the New Hampshire plans were essentially unchanged, Richard Donovan, who had replaced Thomas, rendered a preliminary verdict that the exercise had been nearly flawless.

This premature evaluation came during Donovan's introduction at the public critique of the exercise. The audience was small, as was to be expected at 2:00 PM on Saturday, July 2. The time, Donovan noted "with a straight face," a skeptical columnist wrote, was "most convenient for all concerned." And the situation caricatured itself.

> There was no meltdown. About 1400 officials and volunteers evacuated no people. . . . The organizers . . . called upon no schoolteachers to evacuate the thousands of children who were not in school at the time.
>
> No sunbathers at Hampton Beach were evacuated, either. Astonishingly, no traffic jam resulted.
>
> No citizens in Massachusetts were adversely affected by the nonmeltdown, despite the noncompliance of Gov. Michael S. Dukakis in the nonevacuation.

Richard Strome described the exercise as "demanding, challenging, and fair" and complimented Donovan on his searching, week-long review of the result. Strome's calendar puzzled a *Boston Globe* columnist, who noted that Donovan could apparently compress a week's work into three days.

Neither Donovan nor Strome left the session unchallenged. Diane Dunfey, a Seabrook schoolteacher and a prominent member of the Clamshell Alliance, asked Strome how, on the basis of this drill, the public could be confident of the safety of schoolchildren. Because, Strome replied, "There was simulated placement of school children on buses." And Herbert Moyer, an Exeter selectman and one of SAPL's most dedicated members, wondered why Exeter officials had not been consulted about the date and time of the meeting. Because, Donovan responded, FEMA did "not recognize local governments." That, the *Globe*'s columnist thought, went "a long way toward explaining why FEMA seems so sanguine about results" that he thought ludicrous.[34]

Robert Pollard, a nuclear engineer associated with the Union of Concerned Scientists, submitted a less sarcastic criticism of the exercise. Inspectors from the NRC had noted that plant operators had "displayed questionable," though not disqualifying, "engineering judgement" in controlling one phase of the hypothetical accident. Pollard, however, believed those operators constituted "an inadequately trained onsite response staff," lacking the "ability to develop potential solutions for placing the reactor in a safe stable condition." He conceded that Yankee's operators in this instance had done "no additional harm," but there might be "negative consequences for the public" in the future because the staff had no detailed and informed grasp of plant design and function.[35]

Peter Dame was no engineer. He was only supposed to drive a bus during the drill, and he refused. "I'm not a no-nuker," he said; "If it's safe, plug it in." But he was "pro-evacuation," and he had driven his fourteen-ton empty bus for the last time over a bridge with a ten-ton weight limit. His maps were inadequate, and the plans were "a joke . . . around the bus barn." "The plan, as it is," Peter Dame said, "won't work. It's not safe."[36]

Dame's commonsense criticism—buses ought not to travel over inadequate bridges and drivers ought to have accurate maps—had no effect whatever on New Hampshire's plans, FEMA's exercise, or the NRC's licensing process. Dame's employer said he did not have to drive, which was certainly less troublesome than reinforcing bridges and less expensive than purchasing lighter buses. Yet Dame's brief, tangential involvement in the interminable and complex struggle to operate Seabrook Station is suggestive. Intervenors—local people, usually, with local concerns—were sometimes unconsciously perceived by national officials as defenders of narrow, personal, and parochial interests, whereas the federal government, conversely, was objective and represented larger, more important, national concerns. In this frame of mind, a bus driver knows less about evacuation routes than someone in an office in Washington or in the state capital, and a local selectman or state representative less accurately predicts the behavior of her constituents than a professor who never met any of them.

They were not talking about Peter Dame or buses, but a colloquy among the lawyers in Ivan Smith's proceedings illustrates this gap in perspective. At issue was the availability of shelter at the beach, and Beverly Hollingworth, who represented part of the resort in the New Hampshire legislature, wanted to tell the ASLB that those who owned buildings the state had identified as potential shelters had neither been asked nor given permission for that use. She had talked with owners, visited the buildings, walked the streets, and knew more about actual conditions, admitted the assistant attorney general of Massachusetts, than did expert witnesses that he and his legal adversaries were going to present. Why, asked a frustrated Ellyn Weiss, attorney for the New England Coalition on Nuclear Pollution, "should we decide this case on the basis of what two experts say and not [on] the basis of [what] real people say?" It was not, replied Sherwin Turk of the NRC legal staff, "a question of expertise" but rather "a question of reliability of the evidence." To lawyers, there was no doubt a distinction between "expertise" and "reliability," and however "expert" Beverly Hollingworth was, to NRC lawyers she was "unreliable." In practical terms, therefore, the distinction was without a difference. And thus the Hollingworths and the Peter Dames and the owners of buildings at Hampton Beach could not testify or were excused from driving and did not influence the outcome federal bureaucrats preferred. [37]

That was not the way Live Free or Die, town-meeting democracy was supposed to be conducted in New England, where home rule was more than a slogan. The patronizing, "we know better" attitude of those who supported Seabrook Station unquestionably fed the anger that kept protesters at the gates and attorneys in adjudicatory hearings. For some opponents, the nation's system of governance was at issue, not a mere nuclear power plant. New Hampshire's citizens, themselves participants in town affairs, with easy access to their large band of state legislators, were accustomed to being taken seriously. Some of them thought a rigid, remote, insensitive, unintelligent, paper-shuffling federal bureaucracy, paid with their dollars, was asking them to swallow a $6 billion dose of dangerous technology that somebody who lived, worked, and played elsewhere said was both safe and in the national interest.

Those who articulated this protest never described the process that was supposed to hear them as fair and increasingly used the adjective "rigged." When Daniel Head, who had chaired the first Seabrook ASLB, appeared intrigued by intervenors' arguments, he received an offer he could not refuse. When John McGlennon, an EPA official, disapproved Seabrook's cooling system, his decision was reversed and he went into another line of work. When Helen Hoyt, another chairman of another Seabrook board, seemed prone to error that might lead to judicial intervention, she became ill and was replaced. And when

Edward Thomas could not testify that emergency plans adequately assured the public's health and safety, other FEMA functionaries did.

Not only did the people change, but the rules did as well. In particular, the rule on emergency planning, adopted after Three Mile Island, was repeatedly modified—opponents said "weakened"—through interpretation and amendment as the pressure to license Seabrook and Shoreham increased. If states and local governments would not provide emergency plans, then the utilities or the federal government would do so. If emergency plans were not in place before low-power operation, that was no impediment because low-power operation was not hazardous anyhow. If the system for notifying the public of a nuclear emergency was not ready before low-power testing, that requirement need not be met until full-power operation. If a utility went bankrupt, that condition posed no hazard to the public. When hurdles like those could be cleared, fourteen-ton buses could without risk roll over bridges with ten-ton limits, experienced drivers would have maps, absentee owners would open their beach homes to transients, schoolchildren could be simulated, and accidents would never be "incredible." After careful review during July and August, FEMA announced that the emergency plans for Seabrook Station passed muster.[38]

At the end of the year, the Atomic Safety and Licensing Board concurred. Ivan Smith carefully summarized the lengthy proceedings and then concluded, on virtually every point, that the evidence presented by the state, the utility, and the experts they sponsored was more persuasive than that submitted by intervenors. Minor inconsistencies appeared and unimportant problems were disclosed, but none disqualified and all were susceptible to correction in future versions of the NHRERP. Where evidence was lacking, as in the case of staff requirements at nursing homes, for instance, common sense (though not, of course, the common sense of Peter Dame or intervenors) would serve: "common sense," Smith wrote, "dictates that such considerations are part of a nursing home's licensing requirements." Where large numbers complicated the calculation of evacuation time, the board "intuitively regard[ed] the figures as excessive." Where numbers did not add up, the board did "not believe . . . the Commission's . . . standards" required "the level of mathematical precision" intervenors sought. Where testimony suggested an unplanned contingency, such as a second shift at a reception center, the board found the witness "outside the area of his expertise and entitled to little weight." Emergency planning, in the board's view, was an inexact science—"the qualitative end product of a series of predictions about future unknowns quantified through the use of several assumptions about the time, place, and scope of a possible emergency and the number and reaction of the population potentially involved." Persuaded by the weight of applicants' evidence, the board adopted their suggestions about virtually every one of those variables.

To do so, the ASLB had to reject reams of testimony from academic experts who contradicted the applicants' professors, from local officials who contradicted state officials, and from people who swore they would not behave as plans said they would. This last phenomenon perplexed the judges but did not detain them long:

> We must view these responses in the context of an emotionally charged licensing proceeding when such statements of predicted behavior can be used in an effort to defeat such licensing or for the purpose of pressuring one's peers. We therefore give them little weight in our deliberations.

Presumably, the statements of witnesses whom the board believed were made in another, calmer context and had nothing to do with a license for Seabrook Station.

In particular, the board seemed impressed with the testimony of Dennis Mileti, a professor of sociology at Colorado State University and an itinerant witness for utilities as an expert on human behavior in emergencies. Professor Mileti's reassuring message was that an emergency creates "a therapeutic community," in which people assist one another and, if properly informed and instructed, do what responsible officials ask them to do. Thus drivers do not drive recklessly while evacuating, teachers stay with the children in their charge, police and other emergency workers appear promptly for duty, and people provide rides and shelter and assistance to those in need. From the board's point of view, Mileti's thesis had the great advantage of harmony with the NRC's "realism doctrine," based on "200 years of American history," which postulated that public officials would "faithfully discharge their public duties."[39]

Yet, in relying on Mileti, the ASLB had to judge his presentation charitably and overlook a withering cross-examination. A member of the licensing panel some months before had suggested relevant professional literature Mileti had missed; when he testified, he still had not seen it, though he had asked a graduate student to look it up. His habit was to read emergency broadcast texts several times, estimating their effect on different audiences, Mileti said; he had that very morning reviewed in an hour and a quarter fifty-three pages of script in preparation for his testimony. An incredulous attorney for the Commonwealth asked if that expenditure of time suggested great care. Mileti did not think the implication was fair. He had, after all, been reviewing emergency messages "here at Shoreham" for some time. He was corrected. "Sorry. Seabrook, for a long time." Well, did those messages tell people to lock their doors and leave the area in the event of an accident? Yes, another witness testifying with Mileti replied. And had Mileti himself not previously said that people at the beach would make their facilities available to stranded tourists? Yes, that was his view. But how could that occur if those people were "gone with their doors locked?" The lawyers ignored the frazzled witness and wrangled about striking his testimony. How about signs to identify shelters, Mileti was asked when the legal interlude concluded. Not necessary, Mileti thought, because those without personal means of transportation would depart in the back seats of others' cars, so shelter would not be needed.

That conclusion rested on what Professor Mileti called "altruism, or what have you; there's the general theory." But it was more than a theory:

> there's a raft of empirical evidence that suggests that the hypothesis I offered . . . is the only prudent hypothesis any scientist would offer including the empirical observations that do exist whether they have been gathered through a scientific method or anecdotal evidence would lead us to conclude.

What Mileti meant, if he meant anything, was that evidence, however collected, confirmed his theory. But it developed, in response to an inquiry from Judge Gustave Linenberger, that Mileti could also "prove" his "theory" by the absence of evidence. No one, Mileti said, had ever observed his altruism concept in a real emergency.

> But the generic theory . . . about people helping one another, manifests itself in a variety of ways and at a higher level of abstraction which is the role of science in society. We can make some conclusions and apply them to empirical observations that haven't occurred yet, and I think that is the most prudent hypothesize [sic] for the future.
>
> The lack of empirical evidence to the contrary, is what gives me confidence about making that conclusion.

Judge Smith and his colleagues found Mileti "well qualified to testify on these issues" and discarded the word of those who predicted how they themselves might behave in the event of a nuclear catastrophe.[40]

Acceptance of the pretentious and confused testimony of an out-of-town sociologist in preference to that of local police chiefs or hospital administrators indicated that Thomas Dignan had made his case. Emergency planning was about protecting populations, not individuals; required a best effort but not necessarily success; and need not account for every peculiarity of a plant's location. Intervenors failed to persuade the ASLB of the importance of site-specific information they believed the NHRERP slighted: personnel resources, facilities, the behavior of real people in a real world. The thirty-one-volume plan seemed to Seabrook's opponents to derive from the need for an operating license, rather than from concern for protecting the public. The scheme was cumbersome and so complex that even the experts who prepared it sometimes needed help to decide what was prescribed. All that paper seemed almost unrelated to the reality many residents of the region knew.

Francesca Chicoine, for instance, set out from her Dover home on Memorial Day to pick up her daughter at Hampton Beach. Her trip, which she ordinarily accomplished in less than two hours, required five because of holiday traffic. "I can not comprehend," she wrote, "how people this past weekend would have been safely evacuated off the beach, and this was not even the height of the season." Phyllis Kehoe, a secretary in a Hampton school, knew what she was supposed to do in a radiological emergency. After she had helped the 320 students leave, she would go home; collect some extra clothes, a toothbrush, checkbook, and credit cards; feed the pets; turn off the lights and her household appliances; lock the windows and doors; and stand out on the street. A bus would eventually pick her up; if the first bus was full, she should wait for another. "And I'll sprout wings and fly away," scoffed another skeptical Hampton resident. Debora Gerrish, who lived in Seabrook, thought the plans seemed "good on paper." But she knew people who worked at Seabrook Station and she "wouldn't trust them with shovelling my driveway."

There were seacoast residents who defended the plant, its construction, its regulators, its safety. But they usually did not directly defend the emergency plans. Seabrook Station, Joyce Heath of Newfields observed, was "so safe because of all the precautions they have to take." Or nuclear accidents were as improbable as the collision of aircraft "over Yankee Stadium in the middle of the World Series," as an engineer from Kingston maintained, echoing a promotional cliché of the nuclear industry. Emergency plans, then, would never be used, and the traffic, the complexity, and the ineptitude to which other seacoast residents pointed was irrelevant; the NHRERP did not have to work.[41]

Judge Smith, of course, did not adopt that view, but he did have a lofty disinterest in site-specific detail that did not fit his reading of the NRC's regulations. From the outset, the Commonwealth of Massachusetts had claimed that it was precisely those site-specific details that made Seabrook an untenable site and were the basis for denial of a license. A difficult site required practical emergency plans, not the hyped reassurance of public relations flacks, a system that would effectively protect people from a radiological release, not metaphors about baseball parks. The argument seemed as sound to James Shannon in 1989 as it had to Michael Dukakis in 1986 and to Jo Ann Shotwell in the 1970s. But it had not prevailed in Judge Smith's court, and, with two new colleagues, he would assess New Hampshire Yankee's plans for Massachusetts beginning early in 1989. He did not have far to look for a precedent.

It seemed like old times: Loud music and antinuclear rhetoric at Hampton Beach, where thousands gathered on a Sunday afternoon in June to show support for those planning civil disobedience the following day at Seabrook Station. Bored policemen mustered out on their day off, fighting mosquitoes and inactivity inside the fence of the still unlicensed nuclear power plant. Affinity groups from the Clamshell Alliance—one called Déjà Vu because it consisted of veterans of previous demonstrations—deploying for an occupation of the site, reminding one another of their commitment to nonviolence and looking forward to reunions in seacoast jails. Confrontations between demonstrators blocking traffic and workers seeking access to the plant or counterdemonstrators shouting that the region needed the jobs Seabrook Station provided. Signs and helicopter-borne banners linking the plant to Satan or Chernobyl or political corruption. Company spokesmen disparaging the protesters as a noisy minority, and Guy Chichester on National Public Radio talking once again about grass-roots control of the nation's energy policy.

The Clams had planned the event for months, unaware that a decision to permit low-power testing late in May 1989 would increase the urgency of the rally and the size of the crowds. Mechanical problems and a brief pause while the courts decided the Commonwealth of Massachusetts had run out of legal ways to block low-power operation delayed the first fission reaction until ten days after the initial wave of opponents had dispersed. One of them, waiting to be booked, promised to return "tomorrow, the next day, a week or a month." Old-time Clams, who counted past Seabrook arrests as badges of honor, looked backward as well.

As did some of the police. "It's been years since I've been back here," mused a lieutenant who had begun his tour of Seabrook duty in 1976. He was, indeed, a little perplexed at his return, because he thought the whole controversy had become "a moot point." The owners of the plant appeared to have overcome legal obstacles, political opposition, regulatory indecision, and corporate bankruptcy; enough was enough. But "I guess it's not," observed Lieutenant David McCarthy.

Indeed not, though some things had changed. The weekend's events, for instance, lacked at least some of the tension and rage of encounters a decade before. Clams, still committed to nonviolent civil disobedience, were older and mellower; police seemed less eager to swing nightsticks and spray gas than their counterparts had once been. Lacking a governor who was an acknowledged cheerleader for the nuclear industry, the event lost some of its ideological edge and became more a matter of trespass than of revolution. Hotheads remained, of course, and some of the more picturesque protesters resembled their predecessors in more than clothing and speech. But they were the distracting bubbles in a new and tamer brew: The rally adhered to a program, and the program included paid advertisements, including one from a stockbroker; speakers representing antinuclear factions more moderate than the Clams shared the platform; there was free gourmet ice cream from Ben and Jerry as well as free yogurt, and the music included urban rap as well as pseudo-country-folk. Opposition to Seabrook Station, in other words, had ceased to be an exotic phenomenon associated with style rebels and had become a part of the mainstream. Those who acted on their opposition were undeniably a minority, as utility publicists maintained; but the minority was no longer tiny or exotic, and, at least in the region around the plant, it was not, in the spring of 1989, a minority at all.[42]

There had been other changes too since the protests of 1976. Most obviously, the excavation and stockpiled steel had become a completed generating station, ready since 1986, its owners claimed, to supply needed electricity to New England. The two reactors, to be sure, had dwindled to one, and the second, incomplete and rusting since 1984, had even been erased from some corporate public relations photographs, though the reality, visible from Hampton Beach and elsewhere, resisted retouching. Demand for power, which had borne little relationship to corporate forecasts in the later 1970s, had begun to grow with the robust New Hampshire economy a decade later. Edward Brown, president of New Hampshire Yankee, frequently recited a litany about the region's need for the electricity that would soon be produced at the nation's safest nuclear plant. Just as routinely, a representative of one or another antinuclear group retorted that conservation and Seabrook-induced rate increases would limit the demand for power and that no nuclear plant was assuredly safe. The facts might be new, but the debate by 1989 had become predictable and tired.

The nuclear industry, which had once promised economical electricity without environmental damage, was a notorious failure. Not only did nuclear power require bankrupting private investment, but evidence mounted that government had mismanaged its own nuclear facilities and would not soon have a safe method of disposing of the industry's radioactive waste. Plants producing crucial components for nuclear weapons had to be closed because of safety violations. A chagrined secretary of energy publicly wondered whether his staff was sufficiently skilled to fulfill the department's assignments. "The chickens have finally come home to roost," said Secretary James Watkins a few months after his appointment in 1989, "and years of inattention" to requirements of safety, environmental protection, and public health were "vividly exposed to public examination . . . almost daily." The Department of Justice and the Federal Bureau of Investigation had a few days before raided a Department of Energy installation where toxic and radioactive waste had apparently been dumped illegally. Citizens in Ohio discovered that federal officials had routinely concealed radioactive emissions that Senator John Glenn referred to as "the mini-Hiroshima near Cincinnati."[43] And the schedules for federally managed dumps in Nevada and New Mexico slipped in a manner reminiscent of those for privately owned generating stations a decade before.

Yet these new circumstances did not occasion a new national policy. Richard Nixon's enthusiastic support for nuclear power echoed in every administration thereafter. Secretary Watkins, embarrassed about his department's performance, nevertheless vowed to do "everything in my power" to prevent Long Island Lighting's generating station at Shoreham from being decommissioned, as the corporation and New York State had agreed to do. "If activists can prevent things from being built," Secretary Watkins said, "I can prevent things from being shut down when it's stupid." He hinted that the Nuclear Regulatory Commission could find a pretext to prevent the agreement from taking effect. A spokesman for Governor Mario Cuomo, who had forced the settlement with the Long Island utility, tartly pointed to the obvious irony of Secretary Watkins giving "instructions on nuclear plants" in view of the federal government's record of "mismanaged, hazardous" facilities. Whatever Shoreham's fate, Secretary Watkins was sure that nuclear power would be "a key element in our future national energy strategy," and he promised reorganization of his bureaucracy "so that the American people know we know how to manage things," a pledge that implied an indictment of his predecessors but no shift in policy.[44]

An optimistic Westinghouse employee even discerned "a renaissance in nuclear energy" about the time George Bush was sworn in. To be sure, the nation's largest reactor manufacturer did not have an order in hand from any American utility, but such orders seemed more likely than at any time since the late 1960s. The efficiency of on-line plants had improved; Congress had renewed the Price–Anderson Act, which provided insurance and a limit on liability; there had been slow progress toward siting and preparation of a nuclear dump. The industry's trade association expected smaller future reactors to have standardized designs, which would lower capital requirements and compress the licensing calendar. The association hoped to reduce to one the two-stage licensing process, which involved lengthy and complex hearings both for a construction permit and for an operating license.[45]

In April 1989, the NRC did in fact adopt new rules incorporating most of the trade association's suggestions. The agency offered to approve in advance locations for nuclear plants, a step that would enable utilities to consider site-specific problems, including emergency planning, before construction. The new rules also facilitated use of standard engineering features and designs and combined the construction and operating license phases of the regulatory process. These changes, said Chairman Lando Zech, constituted "a monumental step forward" that would enable utilities to cut construction time in half. Zech also acknowledged that, so far as he knew, there were no pending orders for new plants to end the de facto ten-year moratorium.[46]

The new procedures were supposed to settle controversies about a plant's site before construction. Contentions focused on environmental protection, safety, the quality of materials, and emergency planning would be litigated before major expenditure; the process that had encouraged building a plant in advance of authorization to operate, so baffling to outside observers, was at last reversed. (As had his judicial brethren more than a decade earlier, Judge James Yacos, wrestling in 1988 with the Seabrook-induced bankruptcy of PSNH, seemed astonished that anyone had "put . . . $5.5 billion into the ground at . . . Seabrook . . . *before* determining" that the plant could legally function.)[47]

The reform, of course, was one that Seabrook intervenors had sought years before in the face of adamant opposition from the NRC. The record abounded with motions to halt construction while the geologic structure of the region was investigated, or the location of the intake for cooling water was fixed, or the matter of cooling towers was resolved. Most importantly, intervenors, including the Commonwealth of Massachusetts, had attempted to halt construction after emergency planning rules were adopted following the accident at Three Mile Island. Both the owners of Seabrook Station and the NRC staff had resolutely opposed all those motions, arguing that investment in construction was a risk owners willingly ran. Martin Malsch, of the NRC's legal staff, admitted that the agency had "resisted" (over the dissent of two commissioners, he might have added) the "very persistent efforts" of intervenors "to litigate emergency planning problems" early in the Shoreham and Seabrook cases. "That," said Malsch, "would no longer be the case."[48]

In effect, the 1989 rule vindicated those who for fifteen years had objected to the NRC's irrational licensing sequence. Yet their victory brought neither celebration nor a practical result, for the revision was part of a governmental effort to revive the comatose industry of which Seabrook and Shoreham had become symbols. Had the NRC modified the licensing procedure before Three Mile Island in 1979, Seabrook Station probably would have been generating electricity well before 1989, at an immense saving to both owners and ratepayers. If the NRC had rewritten the rule after mandating emergency plans, Seabrook Station might well have been scrapped with a wasted investment that would have been embarrassing but not bankrupting. Or, at least as likely, emergency plans might have been developed cooperatively with the Commonwealth of Massachusetts instead of through confrontation. However wise, the revision was notoriously tardy.

Nor was it unique, though other changes bespoke the NRC's effort to counter legal tactics of intervenors more than to institute substantive reform. "The dull thud you heard on Friday," the editor of a New Hampshire weekly announced, "was the sound of the Nuclear Regulatory Commission throwing another rule out the window."

Over the last year, four rule changes have favored Seabrook Station's owners. The message that is being sent to the nuclear power plant's opponents is "You're wasting your time. This thing is getting its license."

The NRC's control of the rule book reminded the editor of the school-yard where "the kids . . . who own the ball" revise "the rules of the game . . . because the rules are causing them to lose." Like schoolyard athletes, the NRC had decided rules "are a pain to have around when you want to get something done." The agency had "been subverted to do the bidding of the pro-nuclear-power-forces" and had been transformed into a "bureaucracy whose workers pull roadblocks out of the way for the nuclear power forces."[49]

But the editorial indictment missed the point, because the NRC would have conceded everything but the tone. In fact, the agency's task was to administer the nuclear component of the nation's energy policy. Because of changes in the price of petroleum and an escalated environmental consciousness, among other events, the task had changed some-what since 1954. But the central responsibility, regardless of party or presidential administration, was to facilitate the private development of nuclear power; the political noise surrounding the industry, with rare exceptions, came from state and local politicians, relied on a variant of the "not in my backyard" argument, and did not in most instances fundamentally challenge national policy. The NRC's rules were the federal government's means to assure the health and safety of the public, but they were supposed to foster, not to stymie, the spreading use of nuclear power. There was, then, considerable truth in the persistent accusation of nuclear opponents that the industry and the regulators were in cahoots.

So intervenors eventually had to ground their opposition on doubt about public safety and stimulate mistrust of the technology. Questions about environmental impact, as early Seabrook intervenors discovered, would be resolved. Challenges to the quality of equipment or con-struction could be met with new hardware and more concrete. Outrage about cost was not the NRC's problem. But assuring the health and safety of the public was explicitly part of the agency's charge, and, especially after the adoption of rules on emergency planning, those assurances became more vulnerable to heightened public fear and the political expression of that fear by local and state authorities. The NRC retained firm control of regulations and the entire licensing process. But the very tightness of that control aroused suspicion that defensive, frequently arrogant bureaucrats must be concealing something. The agency's condescending omniscience in response to what doubtless seemed the silly ideas of amateurs neither informed nor reassured a frightened audience; rather, that posture tended not only to irritate but also to convey closed minds without much concern for the public. Open hearings, intended, among other purposes, to allay apprehension through assured and authoritative presentations, preached mostly to those already converted; opponents thought the NRC's adjudicatory process obscured more hazards than it disclosed. Nor were utilities content, though they tended to prevail, because the NRC's proceedings seemed to them dilatory, inefficient, and expensive.

Regulations cannot effectively prohibit fright, and licensing procedures, which did not pay sufficient attention to that phenomenon, stimulated it. Edward Brown and local defenders of Seabrook Station usually mentioned their dedication to public safety. But they did not dwell on it. As the debate heated over emergency planning, they emphasized instead New England's need for power and the economic consequences of probable shortages. Brown was on the scene, and he knew that a rational discussion of technology, where he had every tactical advantage, might devolve into shrill manifestos based more on fear than on what he took to be fact. Those were arguments he could not win. "The opposition," Edward Brown observed ruefully, "has done a beautiful job of scaring the living hell out of everybody."[50]

And there was little reservoir of trust or goodwill for the NRC to draw on in reply. The doubt engendered by Vietnam and Watergate and examples of governmental prevarication undermined the credibility of agencies that sent out soothing press releases about technology everyone knew could be hazardous. That concern escalated when the Department of Energy had manifestly mismanaged the disposal of nuclear waste and concealed the risk and environmental damage incidental to the production of nuclear weapons. Governmental bureaucrats, in the overdrawn, but plausible, indictment of nuclear opponents, were either the conscripted partners of profit-making corporations or incompetent, whereas intervenors, armed only with truth, fought not only rich utilities but their corrupt and unresponsive governmental allies.[51]

More powerful than truth, whoever possessed it, was the law, because disputes of this sort invariably wind up in court. The voters of Sacramento, California, could close the municipally owned Rancho Seco nuclear power plant through a referendum, but voters elsewhere did not control their electrical supply. The state of New York bought Shoreham in order to shut it down, but imminent NRC approval of Shoreham's license and the state's failure to block operation in the courts had forced Governor Cuomo's hand. Political intervention could not be expected in less prosperous states, where the NRC's rules would deflect interference and the judiciary was likely to prevent it.

Both sides in the Seabrook dispute knew they were headed for federal court, no matter how many rules the NRC changed. That was a test Seabrook's owners expected to win, because courts ordinarily do not meddle in the administrative business of executive agencies unless procedural rights have been abridged. That is, judges will determine whether parties in the case have had an opportunity to present testimony or to cross-examine opposing witnesses but not whether a technology is safe or a beach can be quickly evacuated. The First Circuit Court of Appeals, for example, rebuffed a request of Massachusetts to examine the NRC's directive that state and local officials would follow emergency plans prepared by utilities; that was a presumption, the court held, that the NRC was legally empowered to make. Given the length and complexity of the Seabrook proceedings, Seabrook's owners believed a violation of due process unlikely and judicial intervention improbable. Their confidence rested on the solid precedent of the only atomic energy case the Supreme Court of the United States had decided. [52]

Counsel for intervenors vigorously disagreed, at least when talking with reporters. The NRC had "so grossly violated" its responsibility, charged Paul McEachern, whose habit of political campaigning sometimes overcame his legal discretion, that the decision on emergency planning was surely "going to get reversed." Robert Backus was more cautious: "It's not going to be easy," he said, "[b]ut it will happen." He obviously hoped to find procedural errors, because "the thing has been pretty badly handled in our view." [53]

Yet even if Backus could demonstrate flawed handling to the satisfaction of a court, he would not permanently derail the licensing locomotive. If a court discovered error in the work of one of the NRC's boards, for instance, the usual result would be an order that the offending body provide another, legally correct, proceeding. Backus had successfully challenged an EPA ruling early in the Seabrook case; the agency made procedural adjustments and then again reached the conclusion Backus had hoped to overturn. Courts occasionally interrupted and delayed procedures of an executive agency, and they would insist that it follow its own rules, which was, of course, one reason the NRC changed them. But, as Robert Backus knew, the odds that any court would find a pretext to keep Seabrook Station permanently inoperable were long indeed. The bankruptcy of Seabrook's managing partner gave Seabrook intervenors leverage no other antinuclear group had ever had, and it had not sufficed to stop low-power testing. The ultimate recourse was not the courts but the Congress of the United States. The windmill against which intervenors tilted was not a generating station or an individual utility or even the Nuclear Regulatory Commission; often obscured behind those more immediate, more accessible objects of the scorn of antinuclear activists was the real cause of their distress: the nation's energy policy and the Atomic Energy Act of 1954.

Epilogue

On November 7, 1989, the Atomic Safety and Licensing Appeal Board ordered judges on Ivan Smith's Atomic Safety and Licensing panel to reconsider provisions of the New Hampshire Radiological Emergency Response Plan pertaining to evacuation of handicapped persons and schoolchildren and to sheltering transients at the region's beaches.

On November 13, 1989, Judge Smith's Atomic Safety and Licensing Board released a decision, dated November 9, 1989, that approved New Hampshire emergency plans and those prepared by New Hampshire Yankee for Massachusetts. The ASLB decided that questions remanded two days earlier by the Appeal Board did "not preclude the immediate issuance of a license for Seabrook Station" because changes in New Hampshire's emergency plans could "be readily and promptly taken." Massachusetts Attorney General James Shannon promised to appeal the decision.

On November 20, 1989, the Nuclear Regulatory Commission decided that the commissioners themselves would settle all future licensing disputes over Seabrook Station, an action that preempted other adjudicatory proceedings within the agency.

On December 12, 1989, Representative Peter Kostmayer, chairman of the oversight and investigations subcommittee of the House Interior Committee, wrote NRC chairman Kenneth Carr that the commission seemed "dangerously close to twisting" Congress's intent to assure public protection through emergency planning. The commission's "extraordinary actions" in connection with a license for Seabrook Station and "the steady erosion of safety standards enacted by Congress," Kostmayer said, prompted his inquiry.

On December 15, 1989, in a special session, the New Hampshire legislature agreed to Governor Judd Gregg's proposal authorizing North-east Utilities, a Connecticut utility holding company, to buy bankrupt Public Service Company of New Hampshire for $2.3 billion. Legislative approval included permission for seven annual rate increases of 5.5 percent and a partial repeal of the law barring charges based on construction work in progress.

On March 1, 1990, three commissioners of the Nuclear Regulatory Commission voted to grant an operating license to Seabrook Station: "We see nothing . . . that persuades us that Seabrook cannot be operated safely," said Chairman Carr. He himself "wouldn't worry about it at all," he added, if he lived close to the plant. The commission delayed the effect of the order briefly to permit Attorney General Shannon to seek a court order blocking full-power operation. Never, Shannon remarked, had "a licensing issue . . . been so legally vulnerable." Senator Edward M. Kennedy called the commission's vote "the culmination of a long line of irresponsible 'public-be-damned' decisions by the NRC, . . . a rogue agency that lives by its own set of pro-industry rules." Representative Kostmayer called a hearing to investigate the commission's ruling. A spokesman for the pro-nuclear Council for Energy Awareness predicted that the safe, environmentally responsible operation of Seabrook Station would soon persuade even the plant's opponents of the "benefits to their region and their way of life."

About two dozen demonstrators protested the decision at the NRC's headquarters in Bethesda, Maryland. About thirty of perhaps a hundred people blocking the gates at Seabrook were arrested.

In New York, the common stock of PSNH closed unchanged at $3.50. An analyst for a Wall Street brokerage firm called the NRC's action "more like the closing of a chapter than opening one." William Ellis, president of Northeast Utilities, which was expected soon to complete the purchase of PSNH, told an interviewer that "the facts of life . . . say, Don't do another one." The vice president of another partner agreed that the owners' experience "in building the plant is indicative of why no one will ever do it again without drastic changes in the regulatory system and in financing." Michael Mariotte, of the opposition Nuclear Information and Resource Service mused that "we've lost the battle, but we've won the war. No utility wants to go through this again."

Cost of the single completed reactor at Seabrook Station was reported to total between $6.3 and $6.4 billion.

Notes

This book rests almost entirely on documents the Nuclear Regulatory Commission and its predecessor, the Atomic Energy Commission, collected or produced in conjunction with regulatory and licensing proceedings for Seabrook Station. These files are available in the NRC's public document room in Washington, D.C., and in the local public document room for the Seabrook project, which is maintained by the Exeter (N.H.) Public Library, where I did most of my research. This material now extends beyond shelves to entire stacks and has been tended by several librarians in the course of the protracted Seabrook litigation. The federal government has changed the archival filing system during that time, and documents that might once have reposed in folders labeled "Applicants' Correspondence" may now be designated "Legal and Adjudicatory." Transcripts of meetings, hearings, legal arguments, and telephone conferences are not uniformly part of the "Transcript" file. For more than two years in the early 1980s, the local public document room received virtually everything on microfiche, which is filed chronologically rather than by topic; with occasional exceptions, the paper files that bracket this microfiche are organized topically. The NRC's public document room in Washington will provide computerized lists, by archival subject, of the entire collection, from which researchers may select and order copies. Unless otherwise indicated, public documents cited in these notes are included in the Seabrook collection at the Exeter Public Library. Citations to particular folders, especially those compiled before 1981, are not entirely consistent and sometimes reflect my guess about what missing labels may once have said, an apology that must not be construed as a criticism of dedicated librarians who have had to accommodate a flood of Seabrook material that was never the library's central concern. In some cases, I have omitted archival designations because I could not ascertain them and did not want to misdirect those who may wish to investigate further.

1 / **Introduction**

1. Bermanis to Boyd, with enclosures, Jan. 28, 1977; Knighton to Bermanis, Mar. 9, 1977, Environment file.
2. Public Service Company of New Hampshire, *Annual Report to Stockholders, 1972.*
3. See Donald W. Stever, Jr., *Seabrook and the Nuclear Regulatory Commission* (Hanover, N.H., 1980), chapter 1.
4. Richard L. Meehan, *The Atom and the Fault: Experts, Earthquakes, and Nuclear Power* (Cambridge, Mass., 1984), 64, 71, 82; see also Samuel P. Hays, *Beauty, Health, and Permanence: Environmental Politics in the United States, 1955–1985* (Cambridge, England, 1987), chapter 10.
5. See Daniel Okrent, *Nuclear Reactor Safety: On the History of the Regulatory Process* (Madison, 1981), for an inside look at the deliberations of the ACRS by a long-time member of the committee.
6. Quoted in Stever, *Seabrook*, 92–93, and extensively by intervenors in their filings to various NRC boards.
7. George T. Mazuzan and J. Samuel Walker, *Controlling the Atom: The Beginnings of Nuclear Regulation, 1946–1962* (Berkeley, 1983), 76–77. The remark of Director of Reactor Regulation Harold Denton is in the President's Commission on the Accident at Three Mile Island, *Staff Report on Emergency Preparedness, Emergency Response* (Washington, 1979), 42. The licensing procedure outlined here was in effect during the Seabrook litigation but has subsequently been modified.
8. NRC review of ALAB 366, Mar. 31, 1977, 1977 Hearings folder; see also *Time,* July 17, 1978.
9. NRC, memorandum and order on ALAB 471, June 30, 1978, 1978 Hearings folder.
10. Quoted in *Clamshell Alliance News,* July–Aug. 1978.
11. NECNP to NRC, Mar. 3, 1981, G file.
12. Transcript of ASLB hearing, Sept. 19, 1975, 7682.
13. Transcript of ASLB hearing, Oct. 3, 1986, 1071–4, T file.
14. Stever, *Seabrook,* 163.
15. See, for instance, Elizabeth S. Rolph, *Nuclear Power and the Public Safety* (Lexington, Mass., 1979), chapter 2 passim.
16. Meehan, *The Atom and the Fault,* 14, 40–41, 78, 155. See also Hays, *Beauty,* chapter 10.
17. Transcript of ASLB hearing, May 23, 1977, 12758.
18. Quoted in Harry Wasserman, "The Clamshell Reaction," *Nation,* June 18, 1977, 746.
19. Quoted by I. C. Bupp, "The Nuclear Stalemate," in *Energy Future,* ed. Robert Stobaugh and Daniel Yergin (New York, 1979), 108.

20. See James Cook, "Nuclear Follies," *Forbes*, Feb. 11, 1985; Richard Rudolph and Scott Ridley, *Power Struggle* (New York, 1986), 171.

21. Patrick Jackson to William Anders, Feb. 10, 1976, with enclosure from *Manchester Union-Leader*, Feb. 7, 1976; *Boston Globe*, Sept. 26, 1986.

22. Cleveland to Anders, June 25, 1975, July 11, 1975, 1975 Hearings folder.

23. Quoted in Union of Concerned Scientists, *Safety Second* (Boston, 1985), 64.

24. United States Department of Energy, *Nuclear Power Regulation* (Volume 10 of DOE/EIA - 0201 10) (Washington, 1980), 243–44; see also Mark Hertsgaard, *Nuclear, Inc.* (New York, 1983), 63–64, 258.

25. Markey to Palladino, Feb. 29, 1984; Palladino to Markey, Apr. 26, 1984, U file.

26. President's Commission on the Accident at Three Mile Island, *The Need for Change: The Legacy of Three Mile Island* (Washington, 1979), 21–22.

27. Stever, *Seabrook*, 102; Transcript of ASLB hearing, Nov. 4, 1975, 10074–99; Interview with Guy Chichester, Apr. 2, 1987.

28. Transcript of prehearing conference, May 6, 1982, 148, 152.

29. Transcript of sidebar conference, Aug. 18, 1983.

30. Chichester to Hoyt, Aug. 24, 1983; Rye selectmen to Hoyt, Aug. 31, 1983; Transcript of ASLB hearing, Aug. 1983, 1810.

31. SAPL motion to disqualify, Oct. 7, 1983.

32. PSNH reply to SAPL motion, Oct. 25, 1983; see also correspondence from NECNP, Oct. 5, 1983, Nov. 23, 1983, and Massachusetts attorney general, Oct. 31, 1983.

33. ALAB 751, 757; NRC order, Feb. 14, 1984, G file.

34. G. F. Cole, transcript of ASLB hearing, May 24, 1977, 13083.

35. See, e.g., statement of Robert Backus to an unnamed committee of the United States Senate [1983?], typescript in the collection of Herbert Moyer, Exeter, N.H.

36. Ibid.

37. See, e.g., the argument of New Hampshire assistant attorney general Thomas Kinder, Jan. 7, 1983.

38. See Hoyt's order on Amesbury, Mass., Oct. 7, 1986, G file.

39. Staff motion to disqualify, Feb. 4, 1983; ASLB order, Feb. 16, 1983, Apr. 18, 1983; Hollingworth to ASLB, Feb. 25, 1983; Transcript of third prehearing conference, Apr. 7, 1983.

40. Quoted in Union of Concerned Scientists, *Safety Second*, 66.

41. *New York Times*, May 1, 1987, June 4, 1987, June 19, 1987.

42. Ibid., July 14, 1988.

43. Cook, "Nuclear Follies."

44. *Boston Globe*, Sept. 26, 1986.

2 / The Environment

1. PSNH brief seeking review of ALAB 349, 1976 Hearings folder.
2. Transcript of ASLB hearing, Aug. 27, 1975, 5307–8.
3. Sheldon Meyers to Muntzing, July 5, 1974, Environment file.
4. Barrett to Geckler, Jan. 7, 1974, Environment file.
5. Brownell to Mueller, June 3, 1974; Ellyn Weiss to AEC, June 3, 1974; Testimony of Culliney, June 7, 1974; Newell's critique is enclosed with Stever to Geckler, June 18, 1974, Environment file.
6. DES 3–13; Beckley to AEC, Sept. 24, 1974, Oct. 4, 1974, Environment file.
7. FES, section 5.
8. Transcript of ASLB hearing, 11636.
9. See Salo's examination in transcript of ASLB hearing, 10658, 10663, 10870, 10896, 10907–8; his dissent is attached to the ASLB decision authorizing issuance of a construction permit, 1976 Hearings folder, 208, 211, 224, 225, 230.
10. Transcript of ASLB hearing, Aug. 28, 1975, 5409, 5412.
11. Transcript of prehearing conference, Feb. 12, 1974, 707.
12. DES 5–1; Staff testimony for ASLB, Aug. 25, 1975, 2, 1975 Hearing folder.
13. Correspondence between Corps and NRC, Feb. 2, 1975, May 17, 1976, May 26, 1976, Environment file.
14. Nelson to Tallman, Nov. 4, 1976; Backus cross-examination of Beckley, 1976 Hearings folder.
15. Environmental Report—Operating License, 4.1–2.
16. Quoted in Hertsgaard, *Nuclear, Inc.*, 152.
17. Weinhold to AEC, Sept. 1, 1973, Sept. 7, 1973, Hearing folder.
18. See Weinhold's participation in ASLB hearings, 1975, transcript, 2245ff, 2756, 2821, 3232, 3235.
19. Transcript of ACRS meeting, Aug. 22, 1974, 128; Weinhold to Starostecki, Mar. 18, 1974, 1974 Miscellaneous file; Gray to Weinhold, Nov. 14, 1974, Hearing folder. See also Weinhold's limited appearance statement, Sept. 29, 1986, 229–30 in transcript of ASLB hearing, Transcript file.
20. Weinhold to Walter Butler, Nov. 21, 1975, 1975 Miscellaneous file.
21. Goller to Tallman, Jan. 24, 1974; Haseltine to Goller, Feb. 13, 1974, Applicants' Correspondence file; Transcript of prehearing conference, Mar. 11, 1974, 233.
22. Weinhold to NRC, Apr. 30, 1979, 1979 Hearing folder; Transcript of prehearing conference, Mar. 1, 1974, 186.
23. Pomeroy to Kerr, Apr. 15, 1983; Maxwell to Kerr, Apr. 5, 1983; Transcript of ACRS meeting, Apr. 1983.
24. PSNH brief to ALAB, May 8, 1981, G file; see ALAB 422 for an examina-

tion of Chinnery's theories; Chinnery to Weiss, Oct. 23, 1980, Hearing folder; ALAB 667, Mar. 3, 1982, 39.

25. See FES for the controversy over the transmission route.

26. Goldstein to NRC, Aug. 4, 1983. For an example of Dignan's rejoinders, see his response to South Hampton's petition, Apr. 13, 1982.

27. Transcript of ASLB hearing, May 25, 1977, 13287–13294; Stever, *Seabrook*, 150.

28. Denton to Hendrie, Oct. 12, 1973, filed in 1978 Hearings folder.

29. Ott to Denton, June 10, 1974, filed in 1978 Hearings folder.

30. See, e.g., brief to ASLB, Feb. 11, 1977, 1977 Hearings folder.

31. ALAB 366, Jan. 21, 1977.

32. Weiss to NRC, Feb. 11, 1977, 1977 Hearings folder.

33. NRC order, Mar. 31, 1977.

34. Beckley to Knighton, Apr. 4, 1977, Environment file; Prepared testimony of the NRC staff, May 13, 1977, 1977 Hearings folder.

35. Stever, *Seabrook*, 142; ASLB supplemental initial decision, Nov. 30, 1977, 1977 Hearings folder.

36. Transcript of ASLB hearing 12886, 13039, 13280ff.

37. Transcript of oral argument, Mar. 16, 1978, 170, Transcript file.

38. ALAB 471.

39. Backus brief, June 7, 1978; Dignan reply brief, June 21, 1978; Staff brief, June 14, 1978, 1978 Hearings folder.

40. NRC memorandum and order, June 30, 1978; copy in 1978 Hearings folder.

41. NRC memorandum and order, Aug. 9, 1978; copy in 1978 Supplemental file.

42. Backus to ALAB, Sept. 18, 1978; Mulkey to ALAB, Sept. 27, 1978, 1978 Hearings folder.

43. NRC staff testimony related to alternate sites, Dec. 1978.

44. Transcript of ALAB hearing, Jan. 15–17, 1979, 99, 100, 286, 295, 389, 419.

45. ALAB 458, quoted in PSNH brief, Mar. 6, 1978, 1978 Hearings folder.

46. New Hampshire certificate of site and facility, July 27, 1973.

47. Phillips to Muller, June 13, 1974; Testimony of Gekas and Uhler, prepared for ASLB, 18–19, in 1975 Hearings folder.

48. See, e.g., King to Rowden, Jan. 21, 1977; undated [1976?] PSNH press release in 1976 Hearings folder; Beckley to Geckler, Nov. 15, 1974, Environment file.

49. Environmental Report—Operating License, Sections 1.3–1, 1.1, 4–6.

50. Houthakker to AEC, June 3, 1974; Cherry comments, Dec. 13, 1974; undated Gillen comments, Environment file.

51. Transcript of ASLB hearing, 695off, 7120ff, 11112.

52. Cherry comments, Dec. 13, 1974, Environment file.

3 / The Opposition

1. *New York Times*, May 24, 1954.
2. *Reader's Digest*, Aug. 1955, 21.
3. Rolph, *Nuclear Power*, 51–52, 102.
4. Central Surveys, Inc., "An Opinion Survey Regarding Seabrook Nuclear Plant Site, August 28–September 7, 1972." Unpublished survey for PSNH in the library of the University of New Hampshire, Durham.
5. *Hampton Union*, June 7, 1972.
6. SAPL statement of purpose, May 16, 1969; SAPL *News*, June 1971. I am grateful to Herbert Moyer, of Exeter, N.H., who made available these and other materials from his files.
7. Tingle to sportsmen, n.d. [1973]; Tingle to Moyer, Apr. 26, 1984; SAPL *News*, June, 1971; Fiske to SAPL members, July 5, 1972; Moyer papers.
8. Phil Primack, "The Great Power Struggle," in *Boston Globe Magazine*, Nov. 19, 1972.
9. Quoted in Rolph, *Nuclear Power*, 75; quoted in Richard Curtis and Elizabeth Hogan, *Nuclear Lessons* (Harrisburg, Pa., 1980), 203–4.
10. Transcript of ASLB hearing, Oct. 3, 1986, 1077–79, Transcript file.
11. Quoted in John W. Johnson, *Insuring against Disaster* (Macon, Ga., 1986), 19.
12. Transcript of ASLB hearing, Aug. 20, 1983; transcript of ASLB hearing, Aug. 30, 1986, 575, Transcript file.
13. Text of Curran's argument to State Siting Committee, Moyer papers.
14. Stever, *Seabrook*, 21; SAPL *News*, Oct. 1973.
15. Guy Chichester, president of SAPL in 1975–6, has some of the organization's early files, including a folder of correspondence from 1973–4 that contains letters attempting to settle outstanding accounts with Clark and the SAPL's counsel. I am grateful to Mr. Chichester for access to these materials, which are in his possession in Rye, N.H.
16. SAPL *News*, Jan. 1974.
17. Youngs to Moore, Jan. 17, 1975, Chichester papers; Chichester interview, Apr. 2, 1987; SAPL *News*, Mar. 1975.
18. SAPL press release, Mar. 20, 1976; Chichester interview; Anderson to Chichester, Mar. 31, 1976, Chichester papers.
19. Meiklejohn to SAPL, July 29, 1976, Moyer papers.
20. S. Rice to Chichester, Mar. 30, 1976, Chichester papers; Clamshell Alliance, *Handbook* prepared for 1978 demonstration; Wasserman, "The Clamshell Reaction," 744ff; Harvey Wasserman, "The Clamshell Alliance," *Progressive*, Sept. 1977. 14ff.
21. See various Clamshell Alliance pamphlets in the special collections of the library of the University of New Hampshire and in Guy Chichester's papers.

22. See Wasserman articles cited above, n. 20; Minutes of Clamshell Alliance coordinating committee, Mar. 12, 1977, Mar. 27, 1977, Chichester papers; *Clamshell Alliance News*, July–Aug. 1979.
23. Minutes of Clamshell Alliance coordinating committee, Nov. 12, 1977, Chichester papers.
24. Burness and Galvin to Clams, June 10, 1978, Chichester papers: *New Hampshire Times*, June 7, 1978; *Boston Phoenix*, June 7, 1978; Chichester interview.
25. The evolution of the 1978 demonstration may be traced in the minutes of the coordinating committee in the spring of 1978, Chichester papers; Daniel Riesel to Chichester, Mar. 28, 1978, Chichester papers.
26. Spruce Mountain Affinity Group Manifesto, July 1, 1978; *Red Balloon*, June 24, 1978; Account of meeting of Clams for Democracy, June 24, 1978; all in the collection of Sharon Tracy of Amesbury, Mass.; I am grateful to Ms. Tracy for permission to use her collection.
27. *Boston Phoenix*, May 22, 1979.
28. *National Review*, Feb. 2, 1979, 170, 132.
29. Brideau to Anders, Mar. 11, 1976, Miscellaneous Correspondence file; Gilligan to Rowden, Dec. 10, 1976, 1976 Hearings folder; *National Review*, Feb. 2, 1979, 164; quoted in Hertsgaard, *Nuclear, Inc.*, 179; quoted in Karl Grossman, *Power Crazy* (New York, 1986), 176–77.
30. Milton R. Copulos, *Confrontation at Seabrook* (Washington, 1978), 5, 8, 23–24, 38, 39, passim.
31. Chichester interview.
32. Lewis to SAPL board, Apr. 23, 1980; Moyer papers; SAPL *News*, Winter 1980–1.
33. Based on clippings in an extensive press kit furnished by Stephen Comley, Rowley, Mass.
34. Transcript of ASLB hearing, Aug. 1983, 1928; Transcript of ASLB hearing, Sept. 29, 1986, 218ff, Transcript file.
35. Affidavit of Grafton Burke, Oct. 22, 1986, in possession of Sharon Tracy, Amesbury, Mass.
36. Robert Du Pont, remarks at the Conference on Seabrook Station held at Phillips Exeter Academy, Exeter, N.H., Apr. 16, 1987; Transcript of ASLB hearing, Sept. 1986, 35, 71, 575, 589–90, 1135.

4 / Money and Management

1. Quoted in Ralph Nader and John Abbotts, *The Menace of Atomic Energy* (New York, 1977), 219; Irwin C. Bupp and Jean-Claude Derian, *Light Water: How the Nuclear Dream Dissolved* (New York, 1978), 72–74, 81.
2. Quoted in Daniel Ford, *The Cult of the Atom* (New York, 1982), 182.
3. *Durham Paper*, Oct. 12, 1972.

4. PSNH, *Annual Report* 1972.

5. PSNH Bond Prospectus (1974).

6. Rudolph and Ridley, *Power Struggle*, 198.

7. PSNH Common Stock Prospectus (1974); PSNH Bond Prospectus (1974).

8. Source of Funds Projection, Dec. 3, 1974, Applicants' Supplementary Correspondence file.

9. Response to Weinhold interrogatory 13, Oct. 2, 1974; PUC Order, Dec. 31, 1974, in Applicants' Correspondence file.

10. Jackson testimony prepared for ASLB, Mar. 11, 1975, in Applicants' Supplementary Correspondence file.

11. Transcript of 1975 ASLB hearing, 2067, 2031.

12. Transcript of ASLB hearing, May 1975, 1183, 1214–15, 1239–40, 1265, 1560ff, 1655ff, 1733, 1746.

13. Nelson prepared testimony, 17, 19, 31, 34, Moyer papers; Transcript of ASLB hearing, May 1975, 2130ff, 2300ff.

14. Cook, "Nuclear Follies."

15. Transcript of ASLB hearing, May 1975, 1715; PSNH, *Annual Report* 1986.

16. Thomson to Anders, July 1, 1975, Hearing folder; Stever, *Seabrook*, 118–19; *New Hampshire Times*, Apr. 28, 1976; ALAB 422, 93.

17. ASLB, Construction Permit, June 29, 1976, 25, 150, 151, in Hearing folder.

18. Testimony of Harty and Meyer, June 30, 1977, in Applicants' Supplementary Correspondence file.

19. PUC order, May 25, 1978, in 1978 Applicants' Correspondence file.

20. Transcript of NRC hearing, Nov. 2, 1977, 36, 70, Transcript file; PSNH brief to NRC, Nov. 4, 1977; NECNP brief to NRC, Nov. 14, 1977; Massachusetts brief to NRC, Nov. 9, 1977, in 1977 Hearings folder; Transcript of oral argument to ALAB, Dec. 10, 1976, 242, 245, 257–60.

21. ALAB 422, 75–81, 107–14.

22. NRC memorandum and order, Jan. 6, 1978, in 1978 Hearings folder.

23. PSNH, *Annual Report* 1977; Staff brief to ALAB, July 13, 1976; PSNH brief to ALAB, Aug. 31, 1976; Malsch to ALAB, Sept. 8, 1976, 1976 Hearings folder.

24. Donald D. Holt, "The Nuke that Became a Lethal Political Weapon," *Fortune*, Jan. 15, 1979, 74ff.

25. New Hampshire *Revised Statutes Annotated*, 378:30–a.

26. See, among other items in 1979 Applicants' Supplemental Correspondence file, PSNH Common Stock Prospectus, July 12, 1979; Ruling of Massachusetts Department of Public Utilities, June 28, 1979; Ritscher to Vassallo, June 22, 1979.

27. Undated Harrison affidavit to PUC [Dec., 1979]; Affidavit of Eugene

Meyer, Dec. 3, 1979, in Moyer papers; PSNH, 10K Statement to SEC, 1979.

28. Backus to Denton, Oct. 2, 1979; see also Backus to Denton, Mar. 12, 1979, May 4, 1979, Oct. 17, 1979, in Miscellaneous Correspondence file and Applicants' Correspondence file.

29. Director's decision, Nov. 16, 1979, in Miscellaneous Correspondence file.

30. Denton to Backus, Mar. 5, 1980, Supplemental Hearing folder.

31. PUC Order 79–187, Dec. 28, 1979; PUC Report, Dec. 28, 1979, in Applicants' Supplemental Correspondence file.

32. *Business Week*, Dec. 17, 1979, 34; Apr. 14, 1980, 110; see also PSNH investment prospectuses dated Jan. 22, 1980, and Feb. 20, 1980.

33. PUC DR 81–87, 112, copy in Moyer papers.

34. Denton to Backus, July 6, 1982; Perlis to ASLB, June 15, 1984.

35. Quoted in Herstgaard, *Nuclear, Inc.*, 272.

36. Beckley to Geckler, Mar. 25, 1977, Environment file; PSNH proposed findings for ASLB, June 13, 1977, in 1977 Hearings folder.

37. *Granite State Independence*, Dec. 1975; Testimony of Geoffrey H. Minor, Lynn K. Price, and John Stutz, regarding Seabrook I Costs, prepared for the Vermont Public Service Department, Dec. 31, 1986, table IV–2. This document, cited hereafter as Vermont testimony, is in the collection of the Employee's Legal Project, Amesbury, Mass., and was made available to me by Sharon Tracy.

38. Challenge Consultants, *Study of the Seabrook Project* (Shawnee, Kans., 1986), 1:35; New Hampshire Yankee permitted me to use this multivolume study, and I am grateful for the corporation's cooperation. See also Vermont testimony, 44, table IV–2, and Exhibit 8, which incorporates PSNH Planning and Scheduling Report, Apr. 27, 1981.

39. ASLB, Initial Decision (CP), June 29, 1976, 22, 155, in 1976 Hearings folder.

40. Tallman to O'Leary, Apr. 25, 1974, Applicants' Correspondence file; Transcript of ACRS meeting, Apr. 1, 1983; Vermont testimony, 3–8.

41. President's Commission on the Accident at Three Mile Island, *Staff Report on the Managing Utility and Its Suppliers* (Washington, 1979), 120–21, 145ff.

42. Vermont testimony, 47, 48, 51.

43. The Employee's Legal Project, Amesbury, Mass., collected affidavits of employees critical of construction practices at Seabrook; see those of Michael Goodridge, "C," and "J." See also affidavits about drug and alcohol use collected by Patricia McKee, Esq., of Hampton, N.H., especially the statements of Peter Hanson and "X"; I am grateful to Attorney McKee for access to these documents. See, in addition, John Quinn to Nicholas

Costello, June 28, 1985, copy in G file; DeVincentis to Starostecki, June 3, 1983; and Vermont testimony, Exhibit 9.

44. Vermont testimony, 88; Exhibit 6, 9; Systematic Assessment of Licensee Performance, Aug. 1983; Challenge Study, 1: 57; 2: 196.
45. Vermont testimony, 66, 68–69, 70, Exhibit 6.
46. *Manchester Union-Leader*, Feb. 2, 1984, Feb. 3, 1984.
47. Interview with Douglas Richardson, Apr. 30, 1987, Amesbury, Mass.
48. Vermont testimony, 98; PSNH press release, Mar. 1, 1984.
49. PSNH, *Annual Report 1983, 1984,* and *1985*.
50. Rudolph, *Power Struggle*, 182–83; Denton to NECNP, Oct. 17, 1984, G file; Systematic Assessment of Licensee Performance, Mar. 13, 1985.
51. Based on PSNH proxy statements to stockholders.
52. Cook, "Nuclear Follies," 82–95.

5 / Emergency Planning

1. Transcript ACRS meeting, Aug. 22, 1974, 241–60; ACRS letter quoted in Stever, *Seabrook*, 70.
2. Goller to Tallman, Sept. 19, 1973, Applicants' Correspondence file; Denton to Hendrie, Oct. 12, 1973, in 1978 Hearings folder; Stever, *Seabrook*, 223; Karen Sheldon to ACRS in transcript of ACRS meeting, Aug. 22, 1974, 141; AEC to Stever, Mar. 29, 1974, in 1974 Hearings folder; Haseltine to Goller, Apr. 3, 1974, Applicants' Correspondence file.
3. ASLB, initial decision, 164–73, in 1976 Hearings folder.
4. Ibid., 43; see transcript of ASLB hearings, 3626ff, June 19, 1975, for testimony of Capt. George Iverson about evacuation.
5. New Hampshire briefs to NRC, Sept. 22, 1976, Nov. 24, 1976, in 1976 Hearings folder; Transcript of oral argument to ALAB, Dec. 10, 1976, 266.
6. Transcript of oral argument to ALAB, Dec. 10, 1976, 267; Transcript of oral argument to ALAB, Mar. 8, 1977, 17, 43–47; ALAB 390, 1977 Hearings folder.
7. See Stanley M. Nealey, Barbara D. Melber, and William L. Rankin, *Public Opinion and Nuclear Energy* (Lexington, Mass., 1983), 16–17.
8. President's Commission on the Accident at Three Mile Island, *Staff Report on Emergency Preparedness, Emergency Response* (Washington, 1979), 10, 12, 37; see also Ford, *Cult*, 136; Hertsgaard, *Nuclear, Inc.*, 248.
9. Bupp, *Light Water*, 128; Ford, *Cult*, 79; *Science*, Sept. 1, 1972, 771; Richard E. Webb, *The Accident Hazards of Nuclear Power Plants* (Amherst, Mass., 1976), 4; see also Rolph, *Nuclear Power and Public Safety*, 49.
10. Quoted in Ford, *Cult*, 140.
11. NRC press release, Jan. 19, 1979, 1979 Supplemental Hearing folder.
12. Quoted in Johnson, *Insuring*, 130.

13. Quoted by Gary Hart in remarks to the NRC, Jan. 21, 1987, G file; *Federal Register*, Aug. 19, 1980.

14. PSNH response to Sun Valley petition, June 1982; Lorenz to SAPL board, Feb. 12, 1981, Moyer papers.

15. Based on extensive correspondence in 1979 and 1980 Hearings folders; see also 1979 and 1980 Applicants' Correspondence files and 1980 G file.

16. Shotwell to Denton, Mar. 31, 1981; Denton decision, July 15, 1981, G file; Gilinsky dissent, quoted in Exhibit 1, Massachusetts brief to ALAB, Mar. 24, 1989.

17. Bruce Golden, North Hampton, N.H., Mar. 1987.

18. Gallen to Palladino, Oct. 15, 1981; Palladino to Gallen, Apr. 27, 1982.

19. Transcript of meeting on emergency preparedness, Apr. 29, 1982, 14, 36, 66, 77.

20. See correspondence in G file, Feb.–Apr. 1983.

21. See contentions of NECNP, Massachusetts, SAPL, and various seacoast region towns, 1983 G file.

22. Transcript of ASLB hearing, Aug. 17, 1983.

23. Transcript of ASLB hearing, Aug. 19, 1983, 1531–40.

24. Ibid., 1841, 1902–3, 1908.

25. Breiseth to New Hampshire Civil Defense Office, Aug. 17, 1983; Gavutis and Verge to ASLB, Oct. 16, 1983; Hampton Falls selectmen to Beckley, Nov. 22, 1983.

26. *New York Times*, May 6, 1983; PSNH, *Annual Report 1983*.

27. President's Commission on the Accident at Three Mile Island, *The Need for Change: The Legacy of Three Mile Island* (Washington, 1979), 7; *New York Times*, Aug. 27, 1986.

28. See Union of Concerned Scientists, *Safety Second*, chapter 2 passim.

29. Okrent, *Reactor Safety*, 113–26; 129–33, 242–44, and passim.

30. Transcript of ACRS meeting, Apr. 19, 1983, 276.

31. NECNP brief to ASLB, May 26, 1983; Bisbee to ASLB, May 26, 1983; NRC staff to ASLB, June 15, 1983; ALAB 734.

32. Martin to Derrickson, Oct. 10, 1984, G file.

33. ELP response to NRC special investigation, Apr. 1987, in ELP files, Amesbury, Mass.; Ron Ridenhour, "Is the NRC Licensing Unsafe Nuclear Power Plants?" *Gambit*, Mar. 30, 1985, 17.

34. Grossman, *Power Crazy*, 18, 24.

35. Pickard, Lowe, and Garrick, Inc., *Probabilistic Risk Assessment* (1983), enclosed in Dignan to ASLB, March 1983; also in 1983 G file.

36. Charles Perrow, *Normal Accidents: Living with High-Risk Technology* (New York, 1984), 310, 314, 324–25.

37. Robert M. Rader, *Offsite Emergency Planning for Nuclear Power Plants: A Case of Governmental Gridlock*, copy in 1985 G file.

38. Ridenhour, "Is the NRC Licensing Unsafe . . . Plants?" 19.

39. Quoted in Grossman, *Power Crazy*, 243–44.
40. Hampton selectmen to Sununu, Oct. 29, 1985; Strome to Hampton selectmen, Dec. 6, 1985; Backus to Strome, Jan. 9, 1986; Rye to FEMA, Jan. 7, 1986; Walker to FEMA, Jan. 16, 1986; G file.
41. See contentions filed by Hampton, Hampton Falls, South Hampton, and NECNP in 1986 G file.
42. PSNH reply to contentions of the Massachusetts attorney general, Mar. 5, 1986, G file.
43. Transcript of ALAB oral argument, June 18, 1986, 43ff., T file.
44. Transcript of prehearing conference, Mar. 26, 1986, 2235–42, 2247–52, G file.
45. Thomas to Strome, enclosing review, June 2, 1986; Thomas affidavit, June 11, 1986, G file.
46. Dignan to ASLB, June 16, 1986; FEMA memorandum, June 18, 1986; NRC staff to ASLB, July 7, 1986, G file.
47. See, e.g., Rudman to Zech, Sept. 19, 1986, G file.
48. Hampton to ASLB, June 24, 1986; NECNP to ASLB, July 2, 1986; Sneider to ASLB, July 2, 1986; ASLB ruling July 30, 1986, G file.
49. Markey to Zech, Oct. 10, 1986, Jan. 13, 1987; PSNH petition to ASLB, Dec. 12, 1986. See also Curran to ACRS, Oct. 10, 1986, G file.
50. Backus to ASLB, Dec. 24, 1986, Jan. 27, 1987; Massachusetts to ASLB, Dec. 30, 1986; NECNP to ASLB, Dec. 30, 1986; NRC staff to ASLB, Jan. 5, 1987, Jan. 28, 1987; Bisbee to ASLB, Jan. 26, 1987; Hampton to ASLB, Jan. 26, 1987; Maine attorney general to ASLB, Feb. 6, 1987; NECNP to ASLB, Feb. 3, 1987, G file.
51. SECY 87–35, G file; *New York Times*, Feb. 6, 1987, Feb. 7, 1987.
52. Cuomo to NRC, Feb. 23, 1987, G file; *New York Times*, Feb. 25, 1987.
53. McLoughlin to Chilk, Apr. 28, 1987, G file.
54. Transcript of management meeting, Mar. 18, 1987, May 7, 1987, T file; *Boston Globe*, Dec. 19, 1986; ASLB ruling, Apr. 22, 1987, G file.
55. ASLB order, Oct. 7, 1986, G file.
56. Backus to ALAB, Mar. 27, 1987, G file.
57. Transcript of oral argument to ALAB, Apr. 27, 1987, 105, 174, 190–91, 209–14, T file.
58. ALAB 864, May 1, 1987; ASLB order, May 4, 1987; *Hampton Union*, Sept. 18, 1987.

6 / Conclusion

1. *Manchester Union-Leader*, Sept. 15, 1988; *Foster's Daily Democrat*, Sept. 16, 1988.
2. PSNH, *Annual Report 1985* and *1986*.
3. *Hampton Union*, Jan. 20, 1987, June 30, 1987; *Atlantic News*, Apr. 7,

1987; *Wall Street Journal*, May 25, 1987; *Manchester Union-Leader*, July 24, 1987.

4. *Wall Street Journal*, Sept. 21, 1987; PSNH prospectus, Sept. 18, 1987; *Fortune*, Feb. 15, 1988, 106.

5. *Hampton Union*, Sept. 9, 1987; see also "At Home" (PSNH envelope stuffer), Sept. 1987.

6. *Wall Street Journal*, Aug. 6, 1987; *Portsmouth Herald*, Aug. 26, 1987.

7. Harrison to NRC, Sept. 3, 1987, with enclosure; Appendix 9 to Massachusetts attorney general petition to ALAB, March 7, 1988, G file.

8. *Hampton Union*, editorial, Sept. 25, 1987.

9. *New York Times, Boston Globe, Wall Street Journal, Exeter News-Letter*, all Jan. 29, 1988.

10. Backus to ASLB, Feb. 23, 1988; Berry to ALAB, Nov. 5, 1987, G file.

11. Shannon petition to ALAB, Mar. 7, 1988, 5–7, G file; ALAB 895; CLI 88–7 in 28 NRC 15, 16, 26; 272–73.

12. *Exeter News-Letter*, Aug. 30, 1988; *Foster's Daily Democrat*, Sept. 3, 1988.

13. CLI 88–10, 28 NRC 579.

14. Brief to ALAB, Apr. 13, 1989, G file.

15. *Boston Globe*, Nov. 29, 1988, Oct. 2, 1988; Judge James Yacos, "Memo on exclusivity," enclosed in Applicants' Advice to ALAB, June 27, 1988, G file.

16. *Manchester Union-Leader*, Dec. 22, 1988; *Exeter News-Letter*, Nov. 18, 1988.

17. *Manchester Union-Leader*, Dec. 21, 1988, Jan. 22, 1989; *Foster's Daily Democrat*, Jan. 22, 1989; *New York Times*, Feb. 28, 1989.

18. "Public Service Company of New Hampshire's Proposed Disclosure Statement," filed in United States Bankruptcy Court, Manchester, N.H., March 31, 1989, 205–6.

19. *Boston Globe*, Apr. 14, 1989; Campaign for Ratepayers Rights, *Newsletter*, May 1989.

20. *New York Times*, Oct. 6, 1987; part of this account of the opening day of the ASLB hearings rests on the author's observation.

21. ASLB, Partial Initial Decision, 28 NRC 670 (1988).

22. Text of opening statement, 4, 5; provided to the author by Ron Sher of New Hampshire Yankee.

23. *New Hampshire Radiological Emergency Response Plan* IV, G–16.

24. *Hampton Union*, July 7, 1987.

25. Ibid., June 16, 1987; Shepard to Katner, Apr. 15, 1987, Moyer papers; Hampton motion for summary disposition, referring to Strome affidavit, Mar. 10, 1987, G file.

26. 28 NRC 729 (1988).

27. *Hampton Union*, Nov. 18, 1986, Nov. 21, 1986; Transcript of ASLB hearing, 13235 (May 27, 1988), 13449 (June 14, 1988).
28. Transcript of ASLB hearing, 13396, 13419, 13420, 13489.
29. Ibid., 13526, 12506, 13039; *Boston Globe*, June 6, 1987.
30. Transcript of ASLB hearing, 13853, 12718–25, 12776–77.
31. Ibid., 12557, 13720–22; 13610–11, 13723.
32. SAPL *News*, May 1988; C-10 *Newsletter*, Apr. 1988.
33. *Hampton Union*, June 26, 1987; *New York Times*, Nov. 19, 1988; *Boston Globe*, June 30, 1988; *Foster's Daily Democrat*, June 30, 1988, Nov. 21, 1988.
34. Andrew Merton column, *Boston Globe*, July 10, 1988; the stenographic transcript of the meeting differs somewhat from the press account. See Peterson to Stello, Dec. 14, 1988, G file, in which the transcript is enclosed.
35. Pollard affidavit, Sept. 16, 1988, G file.
36. *Portsmouth Press*, June 28, 1988.
37. Transcript of ASLB hearing, 10817–18.
38. *Exeter News-Letter*, Oct. 11, 1988; *New York Times*, Nov. 21, 1988; *Boston Globe*, Sept. 7, 1988.
39. 28 NRC 667ff (1988); quotations at 698–99, 785, 716, 717, 689, and 728–29.
40. Ibid., 741; for Mileti's testimony and cross-examination, see transcript of ASLB hearing, 10051, 10097–98, 10019, 10050, 10138.
41. *Foster's Daily Democrat*, June 6, 1989; *Boston Globe*, Feb. 12, 1989.
42. *Exeter News-Letter*, June 6, 1989; *New York Times*, June 5, 1989; *Foster's Daily Democrat*, June 5, 1989; "Official Program."
43. *New York Times*, Jan. 24, 1989, June 28, 1989.
44. Ibid., Apr. 28, 1989.
45. *Exeter News-Letter*, Jan. 10, 1989.
46. *New York Times*, Apr. 18, 1989.
47. Judge James Yacos, "Memo on exclusivity," enclosed in Applicants' Advice to ASLB, June 27, 1988, G file.
48. *New York Times*, Mar. 25, 1989.
49. *Exeter News-Letter*, Sept. 20, 1988.
50. Ibid., Apr. 21, 1989.
51. See, for example, the op ed piece of Robert Backus in *Boston Globe*, May 19, 1989.
52. Constance Cook, *Nuclear Power and Legal Advocacy* (Lexington, Mass., 1980), chapter 4; Turk to ASLB, Sept. 9, 1988, with enclosure; 367 *US* 398ff.
53. *Seacoast Sunday*, Jan. 29, 1989.

Index

ACRS. *See* Advisory Committee on Reactor Safeguards

Advisory Committee on Reactor Safeguards (ACRS), 10–11, 44–45, 114, 125–26, 135, 143–44

AEC. *See* Atomic Energy Commission

AFUDC. *See* Allowance for funds used during construction

Ahearne, John, 132

ALAB. *See* Atomic Safety and Licensing Appeal Board

Allowance for funds used during construction (AFUDC), 95, 99–100, 102, 107, 108, 163

American Cancer Society, 49

American Civil Liberties Union (ACLU), 78, 90

American Friends Service Committee, 76

Anders, William, 21, 82

Anderson, Dorothy, 75

ASLB. *See* Atomic Safety and Licensing Board

Asselstine, James, 88, 111, 145

Atomic Energy Act of 1954, 11, 23, 64, 199

Atomic Energy Commission (AEC): and cooling systems, 35; and emergency planning, 127; and licensing procedure, 8, 65, 112; powers of, 11; and promotion of nuclear power, 64, and public opinion, 143; and public participation, 65–66; and Rasmussen Report, 131; and safety, 130; and SAPL, 74

Atomic Safety and Licensing Appeal Board (ALAB): and appeals process, 11; and construction permit, 52; and earthquakes, 46; and emergency planning, 129, 132, 150–51, 159–61; function of, 9, 11; and Helen Hoyt, 159–61; and James Nelson, 103–4; and PSNH financial qualification, 103, 168–69; and site comparison, 54, 56–57

Atomic Safety and Licensing Board (ASLB): and beach evacuation, 185–86; and construction permit, 37, 97; and cooling system, 32; and emergency planning, 127–28, 138–40, 151–52, 153–54, 174–76, 178, 187–90; and John Frysiak, 50; function of, 9, 11; and Frank Graf, 70–71; and Daniel Head, 24, 186; and Dennis Mileti, 188–89; and political pressure, 23–24; and PSNH finances, 98–99, 101; and PSNH financial qualification, 166; and PSNH forecasts of electrical consumption, 61–62; and PSNH management, 114; scientific judgment of, 10; and site comparison, 52, 54; and Edward Thomas, 181; and "zero-power" testing, 159

Audubon Society, 27, 41, 71

Backus, Robert: on alternate energy sources, 61; on bankruptcy, 167–68; challenges to Seabrook, 86; and courts, 198–99; and Thomas Dig-

Backus, Robert (*cont.*)
  nan, 25–26; and emergency planning, 85, 155, 159–60, 175; and
  financial qualifications of PSNH,
  103, 109, 170; hired by Audubon
  Society, 71; and Helen Hoyt, 25–27;
  and David Lessels, 101; and low-
  power testing, 174; and Robert Mer-
  lino, 139; and SAPL, 74; and Sea-
  brook opponents, 26–27; and site
  comparison, 55, 56, 57; withdraws
  temporarily from case, 24
Baker, James B., 116
Barrett, B. E. ("Bud"), 35–36
Bartlett, William, 172
Berlin, Roisman & Kessler, 71, 73
Bermanis, H. L., 3, 4
Berry, Gregory, 160–61, 168
Bisbee, Dana, 144, 151
Bores, Robert, 180
Boston Globe, 6, 69, 184
Boyd, Roger, 4
Bradford, Peter, 134
Brideau, Mrs. Benjamin, 82
Brock, Matthew, 175
Brookhaven National Laboratory, 83,
  155, 158
Brown, Bruce, 69
Brown, Edward, 193, 197
Browns Ferry Nuclear Power Plant, 130
Buck, John, 55, 128
Burke, Grafton, Jr., 90–91
Bush, George, 194

Carney, William, 148
Carson, Johnny, 19
Carter, Jimmy, 3
Chernobyl: and licensing of Seabrook,
  163; and opposition to Seabrook, 17,
  27, 87, 88; and private development
  of nuclear power, 23; and retrofit-
  ting, 143; and safety, 70, 142
Chichester, Guy: and Clamshell Al-
  liance, 76; on costs of Seabrook, 113;
  devises media spots, 90; and emer-
  gency planning, 174, 191; and Helen
  Hoyt, 25, 26, 138–39; on police tac-
  tics, 78; and SAPL, 74–76
Chicoine, Francesca, 190
The China Syndrome, 17
Chinnery, Michael, 45–46
Clamshell Alliance, 76–81; and civil

disobedience, 77, 80–81, 191–92;
  and emergency planning, 184; fi-
  nances of, 79; and labor unions, 77;
  and nonviolence, 79, 191, 192; orga-
  nization of, 76; and Thomas Rath,
  79–80; and SAPL, 84–85; tactics of,
  76–78, 79, 80–81
Clamshell Alliance News, 90
Clark, John, 72, 73
Cleveland, James, 21
Coast Guard, 11, 153
Cole, Sterling, 64
Comley, Stephen, 88–89, 174
Commission on Historic Preservation,
  11
Consolidated Edison, 96
Construction Work in Progress (CWIP),
  101–2, 103, 106–8, 164, 165, 167,
  171
Contentions (legal assertions), 9
Cook, James, 100, 124
Cooling system: Thomas Dignan on,
  38, 54, 153; and earthquakes, 44–45;
  and marine life, 32–33, 35–40, 72;
  and ocean temperature, 32–33, 36,
  39; and plant costs, 98–99, 113; and
  safety, 144; and site comparison, 38,
  51, 52; towers, 34–35, 37, 44, 52,
  53–54, 72; tunnels, 34, 35, 37, 39,
  40, 41, 44; and wetlands, 31, 34, 40,
  71–72
Copulos, Milton, 83–84
Corps of Engineers, 11, 41–42
Cuomo, Mario, 157, 194
Curran, Diane, 27, 72–73
Cushing, Robert ("Rennie"), 81–82,
  90, 174
CWIP. *See* Construction Work in Progress

Dahl, Thomas, 3–4
Dame, Peter, 185–86, 187
Déjà Vu (affinity group), 191
Denton, Harold, 50–51, 111, 134
Department of Energy, 22, 30, 71, 197
Derrickson, William, 119–20, 122,
  158, 159
Dignan, Thomas: on adjudicatory hear-
  ings, 9, 10; and Robert Backus, 25–
  26, 27; and Michael Chinnery, 46;
  on cooling systems, 38, 54; on delay,
  40, 61; on emergency evacuation,

133; on emergency planning, 138, 150–51, 152, 175–76, 190; on FEMA emergency plan, 153–54; on financial qualifications of PSNH, 111; on financing of Seabrook, 102, 103–4, 105; and Daniel Head, 24; hired by PSNH, 71; and Helen Hoyt, 25, 27; and David Lessels, 101; on licensing process, 63, 160; on LPZ, 129; and James Nelson, 99; and Jo Ann Shotwell, 139; on site comparison, 31–32, 51, 55, 57; and Edward Thomas, 181; on transmission lines, 47, 49; and Elizabeth Weinhold, 43–44

Donovan, Richard, 184
Doughty, Jane, 86
Draft Environmental Statement, 35–36
Drexel Burnham Lambert, 164
Duffett, John, 162, 172–73
Dukakis, Michael, 77, 86, 141, 157–58, 177, 178, 191
Dunfey, Diane, 174, 184
DuPont, Robert, 92

Earthquakes, 18, 43–46, 146
Eastern Utilities Associates, 167, 168
Edles, Gary, 150, 160
Eichorn, John, 167
Eisenhower, Dwight D., 64, 65
Emergency planning: ASLB reopens investigation of, 174; and beach evacuation, 125–28, 134, 135, 136, 139, 149–50, 151, 180–81; and beach shelters, 135, 175, 181, 185–86, 189; and Thomas Dignan, 133, 138, 150–51, 152, 153–54, 175–76, 190; and Michael Dukakis, 141, 157–58, 177, 178; FEMA tests New Hampshire plan, 152–53; and full-power license, 156; and "generic issues," 142–43, 144; and Hampton, 149, 151–52, 153, 166, 175, 178; and Hampton Falls, 133, 140, 149, 150; and Kensington, 133, 149, 150, 151; and low-population zone (LPZ), 126–27, 128–29, 132; and low-power testing, 154, 169; and Massachusetts, Commonwealth of, 27, 36, 86–87, 133, 138, 140, 141, 148–49, 152, 153, 154–55, 157–58, 159, 163–64, 177–79, 189, 191,

195, 198; and Newburyport, 136–37; and New Hampshire, State of, 133, 136, 141, 149–50, 151, 152–53, 156, 159, 161, 164, 179, 181, 182, 184–86; and New Hampshire Yankee, 155; and public participation, 147–48, 153, 188–91; and risk assessment, 146–47, 155, 159; rules amended by NRC, 156–59; and Rye, 138, 149, 151–52, 153; and Salisbury, 133; and Seabrook, town of, 133; and South Hampton, 140, 150; and state and local government, 86, 132–35, 140–41, 147–61, 174–75, 177–79, 184, 187, 188; and Edward Thomas, 179–83, 187; and traffic control, 125, 136–38, 149–50, 177; and "zero-power" testing, 159. See also Federal Emergency Management Agency (FEMA); New Hampshire Radiological Emergency Response Plan (NHRERP); Rasmussen Report

Employee's Legal Project, 117, 145
Environment, 31–63. See also Cooling system; Earthquakes; Marine life; Ocean temperature; Transmission lines; Turbidity
Environmental Protection Agency (EPA): and cooling systems, 4, 11, 13, 21, 34–35, 39, 52, 54, 57, 186; on licensing, 33–34, 56; and radiation levels, 177; and water pollution, 11
EPA. See Environmental Protection Agency

Farrar, Michael, 54–55, 57, 103, 104–5
Federal Emergency Management Agency (FEMA): approach to emergency planning, 133; and beach evacuation, 136, 137; evaluation of emergency plan, 158, 160; expertise of, 147; and local officials, 149; and New Hampshire's emergency plans, 164, 176; New Hampshire's emergency plans, test of, 152–53, 156, 184–86; and NRC, 11; and Edward Thomas, 179–83, 187
Federal Energy Regulatory Commission (FERC), 11, 170, 171

Federal Power Commission (FPC), 22, 59
Federal Register, 158
FEMA. *See* Federal Emergency Management Agency
FERC. *See* Federal Energy Regulatory Commission
Fermi Nuclear Power Plant, 17, 130
Field and Stream, 68
Final Environmental Statement (FES), 37
First Court of Appeals, 12, 198
Forbes Magazine, 30, 100, 112, 124
Ford, Gerald, 21, 22
Fortune, 106
Freeman, David, 111–12
Frysiak, John, 24, 50, 61–62

Gallen, Hugh, 85, 106, 107, 114, 135
Geckler, Robert, 49–50, 51, 54
General Accounting Office, 30
General Electric, 10
"Generic" issues, 142–43
Gerrish, Debora, 190
Gilder, George, 82
Gilinsky, Victor, 134
Gillen, William, 61
Glenn, John, 193
Goldstein, Charles, 49
Gossick, Lee, 130
Graf, Frank, 70–71
Grainey, Michael, 128, 129
Great Cedar Swamp, 47–48
Gregg, Judd, 169, 172, 173–74

Hampton, 149, 151–52, 153, 166, 175, 178
Hampton Falls, 133, 140, 149, 150
Hampton Union, 166–67
Harbour, Jerry, 154
Harrington, Michael, 14
Harrison, Robert: and AFUDC, 100; and bankruptcy, 166, 167; called upon to resign, 165; contradictions of, 104, 105; and cost estimates, 113; on financing of Seabrook, 97, 98–99, 101–2, 110, 120; on opponents of Seabrook, 107; and PSNH, 122, 162–63, 164; on UE&C, 119
Harty, William, 102
Head, Daniel, 23–24, 186
Heath, Joyce, 190

Hendrie, Joseph, 50
Heritage Foundation, 83
Hickenlooper, Bourke, 64
Hildreth, Robert, 121
Hodel, Donald, 148
Hollingworth, Beverly, 28–29, 174, 185–86
Houthakker, Hendrick, 60–61
Hoyt, Helen: and emergency planning, 152, 155–56, 159–60, 161, 180; leaves ASLB, 140, 161, 174, 186; and opponents of Seabrook, 23, 24–29, 138–40, 177; and Jo Ann Shotwell, 24–25, 139
Hugel, Max, 165, 166
Humphrey, Gordon, 183

Jackson, Tim, 97
James, Barbara, 91–92
Johnson, Wendell, 115–16
Joint Committee on Atomic Energy, 64
Jones, Irving, 69
Jones, June, 32–33

Kehoe, Phyllis, 190
Kennedy, Richard, 13, 55–56
Kensington, 133, 149, 150, 151
Kerr, William, 45, 125–26
Kidder Peabody and Co., 102

Labor unions, 65, 77
Lessels, David, 100–101
Lessy, Roy, 139
Library of Congress, 22
Licensing process: described, 8–14; reform of, 3, 4, 56, 194–95
Linenberger, Gustave, 189
Lobstermen, on Seabrook Station, 69
Loeb, William, 21
Long Island Lighting, 148
Lovell, Ben, 151
Low population zone (LPZ), 126–29, 132
Low-power testing, 154, 164, 166, 169, 173–74, 187, 191
LPZ. *See* Low population zone
Luebke, Emmeth, 151

MAC. *See* Management Analysis Corporation
McCarthy, David, 192
McCarthy, Joseph, 64, 65

MacDonald, James, 125–26, 135–36, 139
McEachern, Paul, 152, 166, 175, 198
McGlennon, John, 21, 30, 186
McGrory, Mary, 88
McLoughlin, David, 181, 182, 183
McMillan, James, 132
Malsch, Martin, 105–6, 195
Management Analysis Corporation (MAC), 115, 119, 120
Manchester Union-Leader, 6, 21, 120
Mann, Marvin, 100
Marine life, 32–33, 35–40, 42, 72
Markey, Ed, 22, 155, 156, 157
Massachusetts, Commonwealth of: and Seabrook, 27, 36, 110; and site comparison, 52. *See also* Emergency planning, and Massachusetts
Massachusetts Institute of Technology, 45
Massachusetts Municipal Wholesale Electric Company, 168
Meehan, Richard, 9–10, 18
Meiklejohn, Holly, 76
Merck-Abeles, Anne, 86
Merlino, Robert, 139
Merrill, David, 114, 115, 119, 122
Merrill Lynch, 121, 164
Metcalf, Mary, 14–16
Metropolitan Edison, 115
Meyer, Eugene, 102
Mileti, Dennis, 188–90
MIT Report. *See* Rasmussen Report
Morgan Guaranty Bank, 102
Morrison, Roy, 174
Moyer, Herbert, 90, 184
Mulkey, Marcia, 50
Mullavey, Wayne, 81–82

Nadeau, J. P., 152
NAI. *See* Normandeau Associates, Inc.
The Nation, 76
National Environmental Policy Act of 1969 (NEPA): and alternate sites, 13; and cost/benefit analysis, 56; and environment, 11; and EPA, 33–34; and NRC, 57; and site comparison, 49, 52; and Seabrook, 5
National Laboratory at Brookhaven, 83, 155, 158
National Review, 82

NECNP. *See* New England Coalition on Nuclear Pollution
Nelson, James, 99–100, 101, 102, 103–4, 105
NEPA. *See* National Environmental Policy Act of 1969
Newburyport, 136–37
Newell, Arthur, 36
New England Coalition on Nuclear Pollution (NECNP): and delay, 13–14; and emergency planning, 13, 144, 154, 155, 186; and Helen Hoyt, 25; and James Nelson, 99; and nuclear power, 27; and PSNH financial qualifications, 103, 121; tactics of, 84
New England Electric System, 173
New England Power Pool, 59
New Hampshire: conservatism of, 5, 6; takeover of PSNH, 172–74. *See also* Emergency planning, and New Hampshire
New Hampshire Cooperative, 148
New Hampshire Fish and Game Department, 13, 35, 36
New Hampshire Radiological Emergency Response Plan (NHRERP), 174–75, 178, 181, 182, 187, 190, 191
New Hampshire State Police, 127
New Hampshire Supreme Court, 121
New Hampshire Yankee: creation of, 121; and emergency plans, 155, 156, 158, 179, 180; independence from PSNH, 169
NHRERP. *See* New Hampshire Radiological Emergency Response Plan
Nixon, Richard, 17, 20, 193
Normal Accidents, 147
Normandeau Associates, Inc. (NAI), 35, 36–39
Northeast Utilities, 5, 173
Notre Dame Lawyer, 70
NRC. *See* Nuclear Regulatory Commission
Nuclear industry, failure of, 193
Nuclear power: and alternative energy sources, 61; and Congress, 65; economics of, 20; and electric rates, 17; and independent audits, 145–46; and insurance, 65; and mismanagement, 124; and national energy policy, 30,

Nuclear power (*cont.*)
64, 65, 111–12, 154, 183–84, 193–
94, 196; opponents of, 19, 20, 64–
94, 141–43, 147; orders for new
plants, 94; plant costs, 94, 112; private development of, 23, 64, 65; and
public opinion, 17–18, 65, 66, 69–
70, 129–32; and safety, 9, 17, 65,
104–5, 129, 130–33, 146; supporters
of, 82, 111–12, 129, 141–43
Nuclear Regulatory Commission: and
cost/benefit analysis, 38, 58, 59, 62;
creation of, 8; and environmental
studies, 11, 36–37, 41, 44–45, 46,
47–48; function of, 23, 28, 57, 128,
196; and "generic" issues, 142–43,
144; and licensing reform, 12–13,
43, 194–96; and low-power testing,
164, 169; and political pressure, 21,
52; powers of, 8–9, 11; and PSNH's
financial qualifications, 97, 105,
110–11, 114–16; and public relations, 19, 40, 134–35, 143, 147–48,
197; and Rasmussen Report, 131–32;
scientific judgments of, 10, 46, 49;
and site comparison, 49–58, 194–
95; supervision of construction, 144–
46; ties to nuclear industry, 29, 182,
196; and whistleblowers, 116–17,
146
Nuclear waste, 17, 27, 29–30, 65

Ocean temperature, 32–33, 36, 39
Okrent, David, 45, 143–44
Onassis, Aristotle, 40–41
Opposition, 23–29, 64–94; and bankruptcy, 167–68; and conservation,
61; difficulties of, 58–59; and emergency planning, 141–42, 154, 155–
56, 174, 190, 195; and environment,
32, 40–41, 46–47, 195; and fear,
197; finances of, 13–14; and licensing process, 186–87, 195, 199; and
low-power testing, 199; and LPZ,
133; and mainstream, 192; and national energy policy, 199; and
NHRERP, 175, 176–79; and NRC,
196–97, 198; and plant costs, 113;
and PSNH finances, 98, 101, 103,
144, 169; and site comparison, 52–
53, 55, 150; tactical errors, 62; and

Edward Thomas, 182, 183; and
whistleblowers, 116–17, 141, 146;
and "zero-power" testing, 159. *See
also* Audubon Society; Backus,
Robert; Clamshell Alliance; Chichester, Guy; Hollingworth, Beverly;
Metcalf, Mary; New England Coalition on Nuclear Pollution; Seacoast
Anti-Pollution League; Shotwell, Jo
Ann; Union of Concerned Scientists;
Weinhold, Elizabeth

Palladino, Nunzio, 22, 125, 135
Parler, William, 156–57
Pathway 2000, 164
Perrow, Charles, 147
Peterson, Grant, 182–83, 184
Peterson, Walter, 72
Pevear, Roberta, 139
Pollard, Robert, 185
President's Commission on the Acci-
dent at Three Mile Island, 23, 115,
130, 132, 142
Preston, Robert, 89–90
Price-Anderson Act, 29, 65, 194
Priest, Elliot, 95
Prince, William, 119
Probabilistic Risk Assessment, 146–47,
155, 159
PSNH. *See* Public Service Company of
New Hampshire
Public Service Company of New
Hampshire: and bankruptcy, 111,
121, 122, 162–73, 192, 195, 199;
board of directors, 122–23, 166;
bond interest rates, 96, 97, 103, 108,
110, 121; and CWIP, 101–2, 103,
106–8, 164, 165, 167, 171; early
plans for Seabrook, 4–5, 60, 66, 94,
95; financial forecasts of, 95, 96, 99,
112, 162–63; financial qualifications
of, 101, 108, 109, 110–11, 166,
167, 168, 169, 170; forecasts of electrical consumption, 59, 60–61, 71,
72, 101–2, 192–93; and insolvency,
97, 108, 109, 110, 111, 166; management of Seabrook project, 22–23,
114–24, 148, 162–63; nuclear fuel
mortgaged, 108, 109; and public
opinion, 48, 62, 66–67, 134, 163;
and public relations, 19, 41, 90,
107, 165; and Public Utilities Com-

mission, 8, 59, 95, 101–2, 104, 106, 110, 121, 163–64, 165, 171, 173; reliance on petroleum, 96; and reorganization of Seabrook project, 119–20, 121–22, 171–72; share of costs of Seabrook, 95, 98, 110, 168; and site comparison, 36, 49–58, 71–73, 113, 127, 151; and state takeover, 172–74; and stockholders, 14, 122–23, 163, 170; value of stocks, 96, 97, 99, 106, 108, 110, 122; and YAEC, 116

Public Utilities Commission: and construction costs, 85, 106; and corporate pressure, 75; and CWIP, 171; and generating capacity, 59; and PSNH's windfall profits, 173; and rate hikes, 8, 95, 97, 98, 100, 101–2, 104, 109–10, 165

Public Utilities Commission of New York State, 49

PUC. *See* Public Utilities Commission

Pullman-Higgins, 117–18

Rader, Robert, 147–48, 154

Rancho Seco Nuclear Power Plant, 197

Rand Corporation, 70

Rasmussen, Norman, 131–32, 146

Rasmussen Report, 131–32, 146

Rath, Thomas, 79–80

Ray, Dixy Lee, 129

Reactor Safety Study, 131

Reagan, Ronald, 111, 145, 148, 157, 183

Reagan Administration, 111, 145–46, 148

Reed, Robin, 90

Regional Assistance Committee, 179, 180, 182

Roberts, Thomas R., 29

Rosenthal, Alan, 54, 57, 151, 160–61

Roth, William, 30

Rudman, Warren, 183

Rural Electrification Administration, 148

Rye, 138, 149, 151–52, 153

Salisbury, 133

Salo, Ernest, 37–38, 39, 53, 98–99

SAPL. *See* Seacoast Anti-Pollution League

SAPL NEWS, 74

Schaeffer, Janet, 92–93

Scharffenberger, William, 162

Seabrook, town of, 133

"Seabrook I," 121

Seabrook Station: benefits of, 3, 58, 59; completion of, 192; and cost/benefit analysis, 38, 58, 59, 62; early plans for, 4–5, 60, 66, 94, 95, 112; and electric rates, 111; and environment, 31–63 (*see also* Cooling systems; Earthquakes; Marine life; Ocean temperature; Turbidity); and faulty construction, 117–18, 144–45, 171; financing of, 4–5, 94–124, 162, 164; and full-power operation, 187; and low-power testing, 154, 164, 166, 169, 173–74, 187, 191; project size, 95, 98, 119, 120, 121; proponents of, 16, 32, 134–35, 186, 191, 197 (*see also* Opposition); and safety devices, 141–44; and site comparison, 8, 38, 49–58; and "sunk costs," 50–51, 56; and "zero-power" testing, 159. *See also* Emergency planning

Seacoast Anti-Pollution League: and the AEC, 74; beginnings of, 67; and Berlin, Roisman & Kessler, 71; and Guy Chichester, 113; and civil disobedience, 76; and Clamshell Alliance, 84–85; and delay, 84; on electrical consumption, 71; and emergency planning, 133–34, 137, 155, 156, 175; evolution of, 26; finances of, 69, 73–74, 76; and marine life, 36, 41; objectives of, 67–69, 85; and public opinion, 86; on Seabrook reorganization, 121

Securities and Exchange Commission, 108, 164, 170

Shannon, James, 168, 169, 176, 191

Shapp, Milton, 3

Shoreham Nuclear Power Plant: bought by New York State, 197–98; decommissioning of, 193–94; and emergency planning, 148, 158; licensing of, 29, 187; and low-power testing, 154, 168; and whistleblowers, 146

Shotwell, Jo Ann, 24–25, 134, 136–37, 139

Silkwood, Karen, 90

Site comparison, 49–58, 71–73; and beaches, 127, 151; and environment, 34, 38, 50–51, 69, 71–72, 113; establishment of siting committee, 8; and marine life, 36; and Elizabeth Weinhold, 43
Smith, Bob, 174
Smith, Ivan, 174, 176, 177, 181, 185, 187–88, 189, 191
Smuckler, Larry, 173
Sneider, Carol, 154
Society for the Protection of New Hampshire Forests (SPNHF), 27, 41, 47–48, 71
Society for the Protection of the Environment of Southeastern New Hampshire, 48
South Hampton, 140, 150
SPNHF. *See* Society for the Protection of New Hampshire Forests
Stello, Victor, Jr., 156–57, 158, 182
Stever, Donald, 16, 49–50, 61, 71, 128
Strauss, Lewis, 64
Strome, Richard, 149, 178, 184
"Sunk costs," 50–51, 56
Sununu, John: and bankruptcy of PSNH, 164, 167; and emergency planning, 179–80, 183; and FEMA, 182; and Paul McEachern, 152; on PSNH management, 119–20; and PUC, 121; on state takeover of Seabrook, 172; support of, for Seabrook, 89, 119, 149, 183
Supreme Court of the United States, 198

Tallman, William, 97, 106–7, 115, 122
Technology Review, 94
Tennessee Valley Authority, 29, 65, 112
Thomas, Edward, 179–83, 184, 187
Thomson, Meldrim: defeat of, 106–7; on delay, 14, 22, 101, 105; and finances of Seabrook, 106; and gag order on state employees, 100–101; and Daniel Head, 24; and oil refinery, 40; and opponents of Seabrook, 21, 75, 77–78, 79; and PUC, 85; support for Seabrook, 13, 72, 89, 119
Three Mile Island: and emergency planning, 85, 129–30, 156, 187; and licensing process, 12, 29; and managerial inexperience, 115; and nuclear accidents, 147; and opposition to Seabrook, 70, 87, 163; and orders for new reactors, 20; and public opinion, 17, 23, 27, 132, 134; and retrofitting, 143. *See also* President's Commission on Three Mile Island
Tingle, Walter, 68
Tracy, Sharon, 90
Traffic control, 125, 136–38, 149–50, 177
Transmission lines, 47–50, 54
Trawicki, Daniel, 102
Turbidity, 41–42
Turk, Sherwin, 186

UE&C. *See* United Engineers and Constructors
Uhler, Robert, 59
Union of Concerned Scientists, 185
United Engineers and Constructors, 3, 114, 115, 116, 118–19
United Illuminating, 4–5, 168
United States Geological Survey, 11
Urbanik, Thomas, 139

Vermont Public Service Commission, 118
Vermont Yankee, 71

WASH-1400. *See* Rasmussen Report
Washington Legal Foundation, 147
Washington Post, 88
Watkins, James, 193–94
Weinhold, Elizabeth, 43–46
Weiss, Ellyn, 52, 186
Wetlands, 31, 34, 40, 71–72
Whistleblowers, 116–17, 146
Whitman, Martin, 165, 171
Wolfe, Sheldon, 91, 92, 140

Yacos, James, 169, 170–71, 173–74, 195
YAEC. *See* Yankee Atomic Energy Company
Yankee Atomic Energy Company, 114–15, 125, 135

Zech, Lando, 169, 194
"Zero-power" testing, 159